SUZANNE JANTSCHER

Eternal Entanglement

First published by SingingStar Book Press 2022

This novel is entirely a work of fiction. The names, characters and incidents portrayed in it are the work of the author's imagination. Any resemblance to actual persons, living or dead, events or localities is entirely coincidental.

First edition

ISBN: 979-8-9872739-1-3

Editing by Joyce Lamb
Editing by Tiffany Tyer
Cover art by Brandi McCann

This book was professionally typeset on Reedsy.
Find out more at reedsy.com

I dedicate this book to the human spirit.

For my father and mother,
whose love, encouragement and words of wisdom live on in my heart.

Acknowledgement

Thank you to the following people who encouraged and supported me along the way in the writing of this story. My family, Steve, Helen, Will, and Ted Jantscher, Brenda Becker and Elsie Jane Anderson. My friends, Dawn Leety, Carolyn Bosher, Alisa Peterson, Faith Loso, Reglindis Mahler Emerson, David Emerson, Lori Sanctuary and Debbie Conrod.

I give my gratitude to my editors, Joyce Lamb and Tiffany Tyer, and I send a special thanks to Julia Dinsmore and Natalie Batalha, whose vision of the world gave me the inspiration I needed to create this novel.

I

Part One

FIRST MOVEMENT IN THE COSMIC SYMPHONY OF LIFE

Neptune, the Mystic
Spreading idealism yet also confusion and deception

Prologue

The year, 2033

Bolting out the door, I race through the corridors with a strange sense of twisted excitement pushing me on. At last I'm there, standing in front of the largest window in the settlement area. Pressing my face reverently against this peephole to the universe, I gaze out in awe. The storm is finally subsiding after days of swirling red dust, and beneath the hazy scarlet sky, the tallest mountain in our solar system is visible again, towering over our small insignificant structures.

But I don't see any stars! With so few windows here, I forget when it's day and when it's night. Frantically I keep searching—I need to see the stars! Ever since I was a little girl something out there has been beckoning to me. I thought I had made sense of it, yet now all that's changing, and I'm scared and confused.

I've gazed through that window too. Are you afraid of what you've seen there?

The old woman's Russian words shock me.

I saw past the galaxies into the infinity of darkness, and I saw the star thrower. I saw you.

Spinning around, I wildly search the room, but she's not here. *Get out of my head, old woman!* I rub my temples, attempting to calm my nerves and clear my thoughts.

Defiantly, you crossed the rift and claimed the right to forge an evolved future. What you do makes a difference, and for this one, you'll make all the difference.

For which one? Panic swarms in, causing another migraine to attack with a vicious ferocity. The room starts spinning, the vertigo's making me nauseous, my hearing is becoming muffled and my vision's fading. *Agh!* Why is this happening? *Joe, you understand me as no one else does. Help me!*

"Tatiana. Tatiana. Can you hear me?"

A fierce pounding is pummeling my brain. Keeping my eyes closed to block out the glaring light, I silently plead in Russian, "Sing for me, Joe. Only sing very strong. Sing when all the day is black and every star has gone."

"Tatiana. Can you hear me?"

"Sing when all the air is cold and home's a far-off spot. Sing for me when others don't, cause they have long forgot." Pausing briefly, I take in a ragged breath then continue sending my emotional message across the expanse. "Only sing very strong, shine, shine, shine with the song."

The drum line inside my head backs off a little, and I crack open my eyes. I'm back in the clinic, once more lying on a hospital bed looking up into the face of the lead medic.

"Who is Joe?" he asks.

T'fu. I must have been saying my thoughts out loud. "My headaches are geteen vorse," I angrily snipe at him in English. Tears begin streaming down my face, and moments later my body is racked with sobbing. I'm dying, the same way my mama did, and no one here understands!

"Tatiana, do you remember where you are?" the medic questions.

"Yez. I know ... ver I am ... Marz," I declare between sobs.

He tentatively sets a hand lightly on my shoulder. "We have checked the results of all your tests and ruled out other issues. You are having migraines, probably from stress. Try to relax and let your body rest," he instructs.

"No!" I hysterically shout as I push aside his hand. "It eez ... more den ... migrainz!" Most of the formative years of my life were devoid of emotions. I felt nothing, until Joe. Only with his support, my mama's diaries and years of learning was I finally able to accept my past and temper the tormented feelings it holds—terrors buried deep within my genetic code. But now something has happened, and all those catastrophic feelings are roaring back with a vengeance.

"Water Bear, Starfish. Water Bear, Starfish." Fervently, I chant Joe's most recent message. His powerful energy instantly envelopes me in a tight cloak of loving warmth, and I soak in its healing powers. Focusing inward, I follow him through our shared emotional memories, relieved that the alarming anxiety is dissipating. "I love you, Joe."

The medic once more places his hand on my shoulder, this time with a firmer grip. "Tatiana, can you tell me what today's date is?"

Why must he bother me with these ridiculous questions?

"Joe is not here on the Mars colony, Tatiana."

You're wrong. Joe is here. Our minds are forever joined and our fates are eternally interwoven. Our lives are forever linked and our souls are eternally entangled. Thank you, Mr. Einstein.

In 1935, Einstein, Podolsky and Rosen discovered an aspect of the universe so strange that Einstein dubbed it "spooky action at a distance." The scientific term is quantum entanglement.

Chapter 1

Venture
Out where the stars live
Where your spirit belongs
Free to sing all its songs
For this is your journey

She ventures
Into the cosmos
Seems she's known all along
That her spirit sings strong
For this is her journey

We all venture
Similar yet different
Drawn by songs without words
Yet the music's still heard
For this is our journey

Tatiana, age six

As I slip into the room, I peek over at the couch. Mama is still lying there. Her headaches are lasting longer now. My brothers blame me.

They say she didn't have headaches before I was born. Slowly, I inch in closer and closer until I am right next to her.

"Mama, Mama." She doesn't move.

I sit on the floor and flatten out the now-crumpled picture that she gave me last week. As I look into it, my pet water bear stares back at me. Papa says we can't have "real" pet bears, but Mama says Tatty is real, she's just too small to see. That's why she gave me the picture. My brothers say Tatty is ugly, but I think she's cute.

"Mama, Tatty and I want to play with you."

"Quiet, Tatiana, let your mama rest." Papa has just walked in. He's home early from work. His footsteps are soft as he comes over to the couch and bends down to give Mama a kiss. Instantly, he bolts up. "Katya! Katya!" he yells. Papa is scared, very scared.

Mama doesn't say anything. She doesn't even flinch at his yelling.

"Why Katya? Why?" Papa shouts even louder.

My head, Papa's pain! Papa's terrible, terrible pain is inside my head, and it won't go away.

Tatiana, age twenty-one
Monday morning, September 17, 2012
Carleton College, Northfield, MN

Do you ever close your eyes then wish with all your heart that when you open them, your life will be different?

I do, all the time, because I live under a veil of isolation through which I view joy and sorrow, sadness and happiness, love and hate from a distance, like a spectator in the nosebleed seats of a stadium. Emotions are undecipherable strangers lurking in the shadows, and I don't think I'll ever know what it feels like to be loved.

In the prolonged silence of the dead of the night, I often spend long moments reflecting. Sometimes, I hear it—the music of the stars.

The faint tune whispers to me and I whisper back, *"Why isn't my soul allowed to sing?"*

My name is Tatiana Pavlovna Novikova, daughter of Pavel Vladimirovich Novikov, ex- commander of the Russian Federation Air Force, and granddaughter of Vladimir Ivanovich Alliluyev, former officer of the Foreign Intelligence Service of the KGB. I am strong, yet I am weak. I am impenetrable, yet I am vulnerable. I am Russian, I am Tatiana, but now I am Anna, Anna Alkaeva in America.

Pizdets! America, the land of new beginnings, the land of opportunity. I don't belong here, still, I want to fit in here. Breathing deeply, I clench and unclench my tight muscles as the insightful, inciteful song lyrics I heard earlier in the day echo in my ears. *Needed to escape there, so I ended up here. Why can't you understaaand meeee? Why is it so—*

"We're skipping ahead to page seventy-two, platonic solids," the instructor announces, effectively shooting down my melodic reverie and plummeting me back to Principles of Chemistry, Mudd Hall, Carleton College, Northfield, Minnesota. And this is where, at the beginning of my sophomore year—which is already dramatically different than my first year of college in Russia—I first encountered Joseph Earl Carleton. He sits three rows down and two chairs left of me, so I can't see him clearly, yet I know anyway.

He exists within a world of isolation.

Like the permafrost on the Siberian tundra, he has a layer that's permanently frozen. I, too, exist within a glacial realm. My ice age formed early in my childhood, right after my mama's death, when my father's raging despair inexplicably crashed into me like a tidal wave. His hideous demons became my hideous demons and his excruciating torment became my excruciating torment. Trapped inside his endless agony, barely clinging to reality, I slipped deep into the depths of his living hell. Hitting bottom, I did the only thing I could; I curled up into a tiny ball and completely shut myself off from everything.

My disconnect was absolute.

Minute blended into hour, and hour into day; all the colors of life were fading away. Then through the vastness of the suffocating space surrounding me, I somehow felt my mama's presence. *Child, don't give up. Fight back, take control.* Valiantly I tried, but my father's grief was too powerful. This world was spiraling away, and I was becoming lost to another.

Time stretched out ... time seemed to stand still ... time ceased to exist.

I have no idea how long I remained in that solitary state. It could have been days or weeks until the emotional connection with my father was suddenly and inexplicably severed. Cautiously, I peeked out and stared around. Everything looked the same, yet it *felt* different; it felt ... empty. For not only had all of my father's feelings been swept away, but all of mine as well. I'd been stripped of the inner workings that make a person whole and left feeling nothing except for my cramped muscles. My soul could no longer sing.

The only companion that remained with me, my lone friend that survived "the change" was my pet water bear, Tatty. More than likely, it was due to the fact that even in the most frigid caverns of our universe at temperatures of -423°F, water bears survive by slashing their metabolism to one one-hundredth of one percent of normal. In that netherworld, connected yet unconnected to what's around them, they cling to life in conditions never dreamed possible, seemingly dead yet alive. Just like me.

Kaboom! From out of nowhere, wrenching heartache tears into me.

Grief, remorse, regret and guilt pummel me from all sides. My long-dormant emotions, they're back! Like Icarus being drawn toward the sun, I'm hopelessly caught, only I'm being pulled toward a bottomless hellhole. *Nooooo!* I can't live through that again. *Focus, Tatiana, Focus.*

Whorls of English scientific terms spewing from the chemistry instructor are careening past me. Three-dimensional space, *whizzz,* vertex of a polyhedron, *whizzz.* My quasi-translator, Ms. Mizuikova, oblivious to my looming apocalypse, continues vigilantly transcribing the lecture on her computer while every bit of my body tenses up and every ounce of my energy pours into tamping down the terrifying fear. *Don't let the emotions get control of you, Tatiana.*

Fixating my retinas to the white board, I focus on a formula, attempting to block out everything else. Somehow, I manage to make it to the end of class. Unlocking my vision, I frantically scan the room, desperate to find a reason for this cataclysmic upending of my world. My gaze almost instinctively zeroes in on Joe as he quickly gathers his things in a rush to leave. He's three steps gone, when he abruptly stops and twists back.

His eyes, those breathtakingly beautiful blue eyes of Siberia, lock with mine. Motionless, I sit transfixed as his penetrating stare pulls me into the depths of his being. Bewilderment, amazement, trepidation, and most incredible of all, empathy flow through me. Why? How? Why now? It's been over fifteen years since I experienced any emotions.

"Anna!" Ms. Mizuikova's sharp, penetrating voice disrupts the connection. "Why are you still sitting there?"

Because of breathtakingly beautiful Siberian blue eyes. Lake Baikal, nicknamed the Blue Eye of Siberia, is the deepest and oldest freshwater lake in the world. Like Joe, it's remote, mysterious and filled with risky, unexplored regions.

Chapter 2

Einstein's dream. For over thirty years, until his dying breath, Albert Einstein searched for a unified theory of everything. His tenacity isolated him from the people around him, and he wrote, "I have become a lonely old chap who is mainly known because he doesn't wear socks and who is exhibited as a curiosity on special occasions." Einstein was simply ahead of his time.

Joseph Earl Carleton, age twenty-one
Monday morning, September 17, 2012
Carleton College, Northfield, MN

I'm so damned tired of living in a world that doesn't understand me. Why do I have to remember *every* element of *every* part of my life, everything I've seen, heard, smelled, touched, thought of, dreamed of and every damn emotion I've ever felt. I can't forget anything.

Yet according to scientific research, an eidetic, photographic memory is a myth. It doesn't exist. Sure, there are proven cases of grand chess masters who recall every move of a game for weeks afterward, and then there's Akira Haraguchi, who in 2006 recited the first one hundred thousand decimal places of pi over a sixteen-hour period, but researchers haven't found a single person who has a true photographic memory. So who the hell am I to dispute the experts?

I'm just a farm boy from rural Minnesota who was kicked out of high school, a mutant forced to remember every instant of his life with total recall, or if you prefer the technical term my father uses, a hooligan. If only some of my memories would fade over time so I could forget certain things, or at least not recall them with such vivid clarity. *Reality, a poem on the tip of my tongue, that I can't quite remember. Reality, a familiar yet distant form, seen through a veil.* I wish astrophysicist Natalie Batalha's poetic view of the world worked that way for me because I'm stuck agonizing over every single moment of every convoluted screw-up in my life, and my life is royally screwed up. I'd do almost anything to have just one day where I'm not haunted by the past, but that's not gonna happen.

"We're skipping ahead to page seventy-two, platonic solids," the chemistry professor directs.

Sitting here listening to this is pointless; I already know everything he's going to say. Taking out my pen, I start doodling football plays in my notebook, filling up an entire page.

"The electron shell model ..." He drones on and on. "The closest shell to the nucleus—"

Wham! Powerful emotions hack into my hard drive of a mind. Caught off guard, I can't block them and they override my firewalls, spreading their invasive agony and causing my already excruciating guilt to explode. I hastily search my memory files, but I can't find the source. Where is this coming from?

It's her! Anna Alkaeva. From the start of the first day of class, I've been disconcertingly aware of her presence, and now, for some incomprehensible reason, her heartache is ripping through me. It's a direct feed, and the intensity of her suffering is unimaginable. Clenching my jaw and gritting my teeth, I try to fight it. What the heck is happening to me?

I can't function like this. My excruciating guilt is already so all-consuming I barely get through each day, since my brain constantly relives every miserable moment of September twenty-second, 2009. The day Lina disappeared. Only by sticking to a rigid strategy of avoiding close personal relationships have I managed to endure these last three, tormented years. Having Anna's anguish forced on top mine is beyond my coping capabilities. *Get out of my head!*

The second class ends I scramble to the door, but after taking only a few steps, I grind to a halt. Even though all the cells in my head are yelling like King Arthur in Monty Python, "run away, run away, run away", it doesn't matter. I can't do it, I can't race away from her. Turning back, I stare up into Anna's dark brown eyes. They're reflecting and dangerously amplifying my own extreme feelings. As the seconds tick by, I'm becoming trapped in her world.

"Anna! Why are you still sitting there?"

The sharp words cut the stranglehold, and I rush out of the building. Yet even as I flee the scene, her presence remains with me. It's persistent, insistent and alarming, and a knowing dread circles in my gut. Her pain will be cached into my memory banks, forever.

Chapter 3

In London's Highgate Cemetery lies the tomb of Russian exile Alexander Litvinenko, victim of one of the most notorious murder cases in modern history.

Litvinenko was a former Russian Federal Security Service officer who publicly accused the Kremlin of ordering the assassination of Russian oligarch Boris Berezovsky. Shortly after that accusation, during a meeting in London with two current Russian Federal Security Service officers, Litvinenko was fatally poisoned.

The Kremlin's message was delivered. Anyone who speaks out against us, no matter who you are or where you are, we will find you and silence you.

Tatiana
Two years ago, Russia

"Tatiana, he almost killed you." My father's voice is rising, and the blood vessels in his neck are bulging.

"I can take care of myself."

"Yes, right into your grave. That man was seconds away from slitting your throat!"

"I can take care of myself, Father," I restate.

"No, you can't. You don't comprehend how dangerous it is." As he rants on about my lack of sense, he begins theatrically waving his hands in front of my face.

Mutely, I glare at him. There's no point in arguing. No matter what I say, he won't listen.

"Running away is a fool's solution, Tatiana. It solves nothing."

Enough. I will not remain here listening to his tirade, crushed under the excessive security he insists upon for my "protection." I'm leaving. Destination—somewhere else, somewhere far, far away.

Tatiana
Tuesday, September 18 to Saturday, September 22
Carleton College, Northfield, MN

As if on cue, my cell phone vibrates. A text message from my father. *Tatiana, another Russian oligarch has been murdered. Be on your guard at all times.*

Ofiyet'! Why can't I escape this? Like Valentina Tereshkova, the first woman in space, I want to break free of earthly bonds and experience what's unattainable to me here. A life of freedom.

For now though, I'm bound to this planet and can't allow myself foolish pipe dreams since I'm hiding from the Russian regime, and even though my father is the only one who knows where I—Tatiana Pavlovna Novikova—am living, I can't count on it staying that way. Therefore, it's imperative I remain alert at all times, which means that until I can determine what's happening between Joe and me, I have to keep my distance from him.

"How much longer are you going to continue traipsing around in this wilderness?" Ms. Mizuikova grumbles in Russian for the umpteenth time. Ignoring her, I use the translating program on my phone and read another sign along the trail in the arboretum. *#7 Lilac Hill. In 1935, Mr. Brand quit the lilac-selling business and donated* *his remaining stock of 1,500 lilacs to Carleton College.*

"I don't think you should be—"

"Not interested in your opinion, Ms. Mizuikova," I curtly cut her off. Gazing around the hill, I don't spy any lilacs. What killed all of them? My search for lilacs seems as futile as my search for figuring out what's happening to me. Heaving a heavy sigh, I move on.

"Your father said that—"

"Not interested in his opinions either," I shout at her over my shoulder as I veer back toward campus.

"I can't do my job if you won't—"

"Won't what?" I round on her. "Sit in my dorm room all day and listen to you translate textbooks?" I storm down the path and increase my speed, hoping to outpace her annoying griping. At the edge of the arboretum, I change course and step onto a wooden slat bridge with sturdy iron railings that crosses over a narrow moat of water.

"You've already wasted over four hours out here and—"

"Leave me alone!" I furiously yell at her for the millionth time. Why am I getting so angry? Wait a minute, why I am feeling anything at all? Scanning the area, I search for Joe. There's no sign of him.

Perplexed, I proceed onto the island where I'm greeted by another sign. *Circling to the center, the ancient practice of walking a labyrinth is over four thousand years old. Journey to the center of your deepest self, then come round to the world with a broadened understanding of who you are. Walking the labyrinth clears the mind and gives one insight into the throes of life's transitions.*

That sure pinpoints the requirements of my current state of being. Approaching the entrance, I determinedly place one foot in front of the other and fixedly follow its winding pathway, attempting to still my thoughts. Reaching the center, I stop, turn my palms upward and search for insight. Instead of enlightenment, more confusion swirls through me, and instead of answers, the same old tormenting question pokes at me. *Why can't my soul sing?*

Since my early childhood, I've been struggling to rediscover the feelings I lost. Countless days I've devoted listening to pop music CDs my aunt Elena sent from America. Being a music teacher, it was her unique way of staying in touch with me. Over and over, I listened to them, trying to feel what the songs were singing about. Sadness, happiness, jealousy, love. I never felt a thing. Yearning for understanding, I expended thousands of hours observing people, watching them laugh then cry, smile then frown, scream then kiss. Their actions made no sense. This labyrinth isn't helping either.

The irony of it all, after years of being trapped in a bizarre, unemotional, matrix of a multiverse, bound to yet separated from the people around me, I finally smack into a rift, only to discover that the firehose effect of Joe's emotions is even more unmanageable than my previous state of existence. Completely exasperated, I abandon the labyrinth, trudge back over the bridge and hike up the hill toward my dorm with a highly irritated Ms. Mizuikova in tow behind me.

The next day with the precision of a Swiss watch, Joe arrives at chemistry class just as it starts and leaves immediately afterward. On Friday, he does it again, never once glancing in my direction. Despite his avoidance tactics, the bond between us continues to tighten. It's becoming more deep-seated and dangerously jumbling my mind during those precious moments in class when I really need to be totally focusing on chemistry. Ultimately, like the crew of Apollo 13, I concede to myself, "Okay, Houston, we have a problem here."

"What's the problem?" Ms. Mizuikova barks at me in Russian.

What is she complaining about now? I roll my eyes in annoyance.

"What does your problem have to do with Houston?" she demands.

T'fu. I must have been thinking out loud. That hasn't happened since I was a very small child. *Pizdets!* Now I've got another big problem, but I'm not telling her about either of them.

"Nothing," I reply, then direct my concentration toward the more dangerous of the two glitches. Until now, my unique ability to barricade emotions has been one hundred percent effective; I've never even contemplated the possibility of a breach. The ramifications could be catastrophic. I have to determine how Joe is messing with my head. But with his feelings constantly bombarding me, I can't pay attention to anything requiring more than a few brain cells. Therefore, puzzling out my problems during class isn't possible and concentrating on chemistry is completely out of the question. Instead, I turn to studying Joe.

His tawny-blond hair is relatively short and wavy, and his skin is a soft bronze color, probably from spending a lot of time in the sun. When he turns to look at the clock on the wall, I spot two scars marking his moderately chiseled features. As I continue my observation, I note his movements are smooth and effortless, even though his muscles are tensed. He has a muscular build, yet not burly; it's more like the well-defined body of a soldier, one who has the hunter instinct of a predator—a trait I've been conditioned to be wary of. By the end of class, I've gained a solid, mental picture of Joe, but there's a huge, gaping hole in my knowledge of chemistry.

When I'm back in my dorm room, mercifully alone for bit, I open my laptop and log in. On a whim, I click on a Russian website and type in the word "chemistry."

Chemistry, the study of the composition, properties and behavior of matter.

Chemistry, the composition, properties and reactions of a particular substance.

Chemistry, the nature and effects of any complex phenomenon.

Chemistry, a reaction, taken to be instinctual, between two persons.

BINGO! Bingo? *Blin!* Hannah-talk is infiltrating my vocabulary.

"Aaanna." My roommate's high-pitched voice screeches through my head as she blusters into the room. Simultaneously, Hannah's shocking appearance bursts on the scene like glaring headlights on a dark, country road. Her eyelids, lips and closely clipped fingernails, all in brilliant shades of blood red, conspire with her long, green, dangly earrings shaped like parrots and her psychedelically colored, complex mix of clothing styles to blaze an eight-lane highway through my retinas. Her outrageous fashion collage is a peculiar cross between a Russian gypsy and a glamorous movie star. Do all people from small towns in Kansas dress this way? I wish my father hadn't been so effective at drumming in the necessity of meticulously observing everyone around me. All these extraneous details are clogging up my thoughts.

"How was your day?" Her screeching pitch pierces my eardrums again.

Not now, Hannah. In fact, not ever. She's an incredibly irritating thorn in my ear that I thought I had taken care of. With brutal clarity, I had conveyed to her that we are not friends, not comrades, not amigos and certainly not "bosom buddies." As roommates, we share physical space only, nothing more.

"Was it as fine as a frog's hair split four ways?" Her grande-size smile nearly blinds me. Apparently, my limited knowledge of English hindered me in making my point unmistakably clear, for she's still under the misconception that I want to listen to what she's saying. Disregarding her ridiculous question, I switch to staring at my computer screen. How would I respond to that anyway? *My day was utterly disarming, Hannah, it split my frog hairs.*

"You shouldn't worry so much about chemistry, Anna. You're making an elephant out of a fly," she counsels me as she waltzes across the room, her high heels clicking on the floor.

Elephants don't fly. What a nonsensical thing to say.

"I've been trying to learn Russian phrases because I want you to 'be in your own plate' as my roommate. How am I doing?" Out of the corner of my eye, I see her pulling some unusual things out of her backpack: an overly large, green bell pepper which she bites like it's an apple, what appears to be a rabbits foot that's been dyed an obnoxious pink color and a small, stuffed, toy tabby cat she tenderly places on her pillow. "Work is no wolf. It won't run away into the woods, Anna. You need a break."

"No, she doez not." Ms. Mizuikova sweeps into the room, marches directly to Hannah and sternly rebukes her. "She needs du vork even harder."

Wow. For the first time since my father forced her upon me, I'm glad for Ms. Mizuikova's interference. With a sense of satisfaction, I watch her imposing figure stare Hannah down, an amazing feat I've not yet managed to accomplish. Once Hannah retreats, Ms. Mizuikova triumphantly strides over to me, sits down, opens up my chemistry book and starts reading chapter eight in Russian.

I try to pay attention, but images of Joe keep distracting me. While all my other subjects are well within my grasp, chemistry remains out of reach. Joe's close proximity during class completely disrupts my thoughts, and the more effort I exert in steeling myself from his emotions, the more powerful an effect they seem to be having on me. I've got to find a way to put a stop to this.

After what seems like an endless night of mulling over solutions for my *chemistry* problem, Friday night dissolves into the wee hours of Saturday morning. Somehow, I manage short spurts of restless sleep, but my weary brain keeps picking up right where it left off. How is Joe fouling up my systems? Lying in bed, staring up at the ceiling, I analyze it from every angle. Nothing. Every assessment I make is flawed. It's best to step back, take a break and attack the issue later.

Heading out the door, I race through the corridor with a peculiar, twisted energy pushing me on. The walkway seems longer than usual. At last I'm there, pressing my face to the largest window and reverently gazing out at the Martian landscape. But I don't see any stars. Frantically I keep searching. I need to see the stars! Panic swarms in, the room starts spinning and I'm feeling faint. I try to brace myself, but I can't. My head hits with a loud clunk and ... I'm jolted awake.

This is the third time this week I've had this dream. The Mars part I get. Since last year when billionaire Elon Musk announced plans to send humans to the Red Planet within the next decade, I've been waiting for a sign-up sheet. I want to be part of that crew, or any other crew going to space, such as The Mars One mission. But why is this dream so over-the-top emotional, and what am I so frantically searching for? What is my subconscious trying to tell me?

It's definitely time to get up and take a break. I discreetly shift my eyes over to see if Hannah noticed me falling off the bed. If she did, she's not letting on. I get dressed as quietly as possible, wanting to sneak out of the room without disturbing her. As I tiptoe past her bed, the vision of Hannah unexpectedly rising from the dead like the mummified body of Imhotep in a London museum whips through my head. This is outrageous! How am I supposed to live like this? How does anyone live like this, with all these irrelevant emotions muddling their head? Rewind, erase, begin again. I'm almost out of the room when someone down the hallway slams a door, waking Hannah up.

"Anna, is that you?" she sleepily squeaks in a way beyond soprano voice.

No, Hannah, it's just a little fly becoming an elephant.

"What are you doing up so early? You certainly didn't go to bed with the chickens last night."

There's no chance of deciphering that. Shutting the door, I bound down the stairs and out the dorm, then make a beeline for the fine arts building to withdraw into my earthbound refuge, Music. The musical realm allows me to peek into a place that until very recently I've been cut off from, the world of emotions.

Soon I'm holding a fifteen-key Shona kalimba, an African instrument that has existed for thousands of years. Long ago, it was believed traveling storytellers' played its mystical sounds to reach the heavens and call spirits down to Earth. Who knows, maybe it can bring me some off-world intervention. Making my way to a practice room, I sit down and begin rhythmically plucking its tines. At first, there are only in-harmonic tones, however, as these fade away almost-pure tones remain. The ethereal timber entrances me, and I spend most of the morning under its spell.

Regrettably, my reprieve is short-lived, for I have to return to my room to endure another tedious study session. With leaden feet, I sullenly truck across campus, slog up the stairs and hesitantly open the door. Hannah's not here. Ms. Mizuikova's not here. *Ypa!*

Five minutes is all I get before Ms. Mizuikova strides in. Nonetheless, in that brief time three things jumped out at me. Direct recon on the source of my dysfunction—Joe—is needed, venturing out into the theater of operations is necessary and direct observation is required.

Unfortunately, I have no idea where Joe hangs out, but I bet Hannah does. While I don't want to do anything to encourage her chummy behavior, I don't see a way around it. Any plan to seek Joe out will require a bit of assistance from her.

Shortly after my study session ends and Ms. Mizuikova departs, Hannah and all her carbonized effervescence bubbles into the room. She's learning my schedule and is getting quite adept at avoiding Ms. Mizuikova. Unsurprisingly, the two don't get along.

Turning toward her, I inwardly cringe and utter, "I vant du go out tonite."

Hannah's eyes widen in shock for a brief second before she recovers from the bombshell I lobbed at her. "Well, shut my mouth, I do declare. Where do you wanna go, Anna? What do you wanna do? Who shall we invite to go with us? I'm so turnt about this! How long do you wanna to stay out? And what should we wear?" She's drooling excitement.

What have I done?

Chapter 4

Tatiana
Saturday, September 22

That evening, I'm strategically seated with a line of sight to both entrances and my back to the wall. Hannah, three of her friends and Ms. Mizuikova are surrounding me at a circular table in the Contented Cow Pub. I immediately scope out everyone inside, analyzing their threat level. Two individuals catch my eye. One is seated in a large group, wearing military fatigues and blatantly staring at me, and the other is alone, wearing sunglasses and looking away from me.

But no one stands out as much as Hannah. Her current choice of colors is differing shades of neon orange and iridescent purple. Thank the stars there are no black lights in here, or she'd be glowing so brightly I wouldn't be able to see anything else.

I'm able to avoid direct communication with Hannah and her entourage by occasionally glancing in their direction and nodding my head. While sipping my pop, I continue discreetly scanning the room and monitoring the doorways. All of a sudden, my body clenches of its own accord, and Joe enters through the upper door. As he walks toward the bar, everything in my periphery recedes, and I sit motionless, zoning in on him through my tunnel vision.

"Aaaana." A far-off squeal wiggles into the outer reaches of my cognizant self. "Yoo-hoo, Aaaana!"

Painfully, I drag my attention back to the circular table we're seated at. The eyes of every member of the Hannah Sisterhood are riveted on me.

"He's no good for you, Anna. He isn't worth two cents and won't amount to a hill of beans. He's trouble with a capital T, and I'm amazed he hasn't been kicked out of Carleton yet." Hannah declares this in the ultra-upper octave of her shrill voice, ensuring that I hear every word of her wisdom over the pub noise.

"Fo shizzle, Hannah! Even though Carleton's jacked, and it kills me to diss one of the hottest guys on campus, you're spot-on. He's toxic with a capital T," a fellow member of the sisterhood chirps in her agreement.

Yolki-palki! Their unintelligible slang is scattering chaff through my mind and blocking my synapses. Why can't they speak in English? This additional language barrier will further prevent me from understanding their answers if I decide to interrogate them for relevant intel. Oh well, that's not a viable option now anyway, not with Ms. Mizuikova planted beside me. I don't want her getting involved in my personal life any more than she already is. I'll simply question Hannah later, when we have more privacy.

"Anna, you need to steer clear of him." Reaching over like a nanny, Hannah pats my hand in three, short, staccato beats as she reasserts her advice. "One of my friends was in here last year when Carleton got arrested for fighting. She said he totally lost his cool and kept pounding some dude until a few other football players pulled him off. He's a bad apple, and you shouldn't go near him with a *fifty-foot* pole." She places extreme emphasis on the "fifty-foot."

What does a fifty-foot pole have to do with going near rotten apples? I certainly wouldn't carry something like that around an orchard to measure things, and it would take forever for me to convert each measurement to the metric system. *Ugh.* Why am I bothering to try and understand her kooky sayings?

Enough sitting around here doing nothing, I have to get nearer to Joe. The moment opportunity presents itself, Ms. Mizuikova leaves to find the bathroom, I take advantage and quickly wind my way over to the bar. To blend in, I place an order with the bartender. Instead of getting my drink, he stares at me with a deer-in-the-headlights look. "I vant Kvass," I loudly repeat, but the deer continues to blindly gaze at me.

All this while, the link between Joe and me is palpably buzzing like electricity around high-power electric lines, making my entire body hum with an agitating, nervous energy. Oddly, it's rather exciting, and I find myself strangely attracted to him. Stealing a sideways peep, I attempt to see if he's feeling it too ... I think so. His head is lowered and his fingers are tightly clutched around his drink.

Returning my attention to the bartender, I give up on what I really want. "Cola-pop!" I yell. My outburst jolts the guy out of his daze and prompts him into action. As he turns to get my drink, I unleash a litany of choice Russian words.

Joe abruptly coughs. Not an I-have-a-scratchy-throat cough, but a sputtering, I'm-shocked-and-inhaled-some-of-my-drink cough. Taking another glance, I notice he's intensely concentrating on the countertop, purposely avoiding eye contact. That's when it hits me. "You understood what I said!" I spurt out in Russian.

He remains still as stone, not moving a millimeter.

"Has the cat bit your tongue?" I prod him with one of the many idioms Hannah uses on me. No luck. All right, back to my tactics. "Look at me." He steadfastly maintains the theory that I don't exist. "I know you understand Russian, or you would've told me to speak English by now."

Finally, there's a small dent in his armor; his jaw twitches. A moment later, he slowly tips up his head, and his hauntingly blue eyes figuratively and literally bolt me in place. "Talk in English," he grunts.

His harsh retort triggers a bitter fury in me. "No, you talk in Russian!" I indignantly snap back while smacking my hand loudly on the surface of the bar. It seems my temper has been growing over the years it was cloistered away.

Joe immediately returns his focus to countertop. Inconceivably, this small act of rejection shatters my internal defenses, allowing the full force of his unchecked emotions to rush into me. There's agony, torturing self-condemnation and Draconian guilt. I have no idea how to handle this.

"Haven't you ever felt the need to just talk to someone?" I spit out. Did I really say that? I can't confide in anyone. It's not safe. Since my earliest years, my only confidant has been myself. For when you share your innermost thoughts, they're no longer your own. They attain a separate existence beyond your grasp and stroll down paths you don't want them to take. Some become embellished and claimed by others. Some are shuffled along undetected until a *chut-chut,* the tiniest bit, calls attention, and you're called out. The remaining few are sheltered and pampered, only to be brutally ripped out by their roots.

"Don't you see the devastation your refusal is wracking upon me?" I exclaim. How is he doing this to me? Hundreds of people have given me the cold shoulder without any result. The implications of my situation barrel down. I am at this moment completely unprotected and wholly at his mercy. Horrific fear grips me and my eyes tear up. The last time I cried was when my mama died.

Frantically, I struggle to regain control. *Recalculating.* Like a chiding from a first-generation GPS unit, the word ricochets in my head. *Recalculating.* It's accelerating Joe's turbulent feelings as they rip through. *Recalculating.* It's shredding everything within reach and exponentially multiplying the discombobulating mess of tangled sensations I'm experiencing.

A few teardrops escape and roll down my cheek. I can't stay here! Swiftly I twist away, only to terrifyingly realize I have an audience. While I've been living out this nightmarish scene, sequestered in my own little bubble, everyone else in the pub has been watching me. This is exactly the sort of attention I'm supposed to be avoiding. Haphazardly, I plow my way toward the exit, needing to escape before I'm weeping in front of all these strangers.

"I'm sorry." Joe's apology stretches across the room, spinning itself around me with a perilous attraction, like strands of a spider's web capturing its prey. My steps become sticky and it's hard to move.

"Простите." His Russian apology sends the sting that immobilizes me. Seconds later his fingertips brush my upper arm, causing a strange warmth to tingle though my body. "I was only thinking of myself. I didn't want to screw up my life more by talking to you in Russian."

"Then why are you?" I meekly bleat.

"Because I can't endure your pain."

"What? What do you mean?" I'm bewildered by his answer, by my reaction to his touch and by the tears still leaking from my eyes.

Joe wrestles with what to say for so long that I don't think he's going to answer. Looking into my eyes, he finally whispers, "I feel your suffering inside of me, and I can't stop it."

He's feeling my pain? He's experiencing my emotions the same way I'm experiencing his?

"Can we go outside and lose the audience?"

Being reminded of the presence of a room full of observers pushes me to the edge, and I teeter on the precipice, physically losing my balance. Joe slides his arm around my waist and gently guides me out the door, then walks me down the sidewalk. Here, away from everyone, all my barriers completely collapse and violent sobs wrack my body with a deadly force. No amount of steel determination will slow the onslaught of suffering that I've been repressing for fifteen years.

Joe wraps me in a tight embrace and allows me to give into an epic bout of crying. For a long time we stand like this, until at last, my tears runs dry. I'm weak, so weak. I'm also horrified. Trying to push away, I confusedly state, "I'm sorry, I never feel emotions."

"Wish I could do that," he mumbles, then tenderly resettles my head against his shoulder. Without saying another word, he's able to ease my suffering more than all the people who descended upon me with their incessant chattering after my mama's death. I relax against his warm chest and begin to wonder. "What did you feel? When you said you felt my suffering inside of you, what was it like?"

"Saudade," he instantly responds. "An all-consuming melancholy longing for someone you love who is lost, along with a repressed awareness that the person you desperately yearn for will never return. You're left with an agonizing ache for something that does not and probably cannot exist ever again." His words transport me back to my childhood home where my younger self is kneeling beside the couch. In my mind's eye, I see my father leaning over my mama. His face is gruesomely distorted with agony.

"What causes you so much pain?" Joe carefully brushes a tuft of hair out of my eyes.

I can't tell him that. I can't reveal anything about myself to anyone ... but what will happen to me if I don't talk to someone about this? It's obvious I can no longer keep the repressed trauma locked inside of me. I'm caught in a complex web that's both stretching me away and pulling me closer to Joe.

The intense feeling of saudade returns, but this time, it's slightly different. It's not my father's pain, it's Joe's. Reaching up, I trace my finger along one of his scars. "What causes your suffering?" I ask.

His anguish deepens, and I acutely feel every iota of his torment. Without conscious thought I lean in, inadvertently putting our faces a hair's breadth apart. His enticingly warm breath skims across my cheeks. Then a moment later, his heated lips caress mine, sending an unimaginable sweetness through me. For the first time in a long, long time, a semblance of joy radiates within me, and I huddle in the wonderful sensation.

Deepening the kiss, Joe covers my mouth with a sumptuous intensity. Our tongues intertwine, our souls join and that huge, icy void within me is shoved aside. Miraculously, I experience a blissful peace where there are no dangers, no fears and no pain. There's only this mysterious oasis and the allure to remain here forever.

Abruptly, Joe pulls back with a sudden jerk. The swift break allows the outside world to regain a hold on us, and the buildings of downtown Northfield begin to rematerialize. Breathlessly and wordlessly, we stare at each other, transfixed while the rational regions of our brains reboot. With sadness, I watch as foreboding waves of regret cloud over Joe's brilliant blue eyes.

"I shouldn't have done that, Anna. It's only going to cause both of us more misery."

More misery? I've just experienced a sense of acceptance, understanding and compassion that has been woefully lacking in my life, and he's labeling it as misery?

Hooking his arm through mine, Joe ushers me back to the entrance of the pub with a grim look molded on his features. "I don't want to hurt you, Anna. Go back to your friends before they call campus security on me." Then without coming to a full stop, he quickly pivots and walks away, leaving me standing there, confused, lonely and ... missing his touch.

Chapter 5

A human being is a part of the whole called by us a universe, a part limited in time and space. He experiences himself, his thoughts and feelings as something separated from the rest, a kind of optical delusion of his consciousness. This delusion is a kind of prison for us, restricting us to our personal desires and to affection for a few persons nearest to us. Our task must be to free ourselves from this prison ...

Albert Einstein

Joe
Saturday evening, September 22

I'm majorly messed up by my ever-present guilt as I enter the Contented Cow pub. It's siphoning out my soul like a dementor from Harry Potter. This day sucks! Every September twenty-second sucks. After three long years, the disappearance of my girlfriend, Lina, still brutally rips me apart, and the memories of her continue to punish me with the same vicious intensity. There's no solution for it, only endurance.

Walking up to the bar, I order a drink. *Shit!* I feel Anna's grinding tension from across the room. The last thing I need tonight is a battle with her and her twisted issues, I've got nothing left; I'm already wasted from fighting off all my own demons.

Taking a large swallow of my beer, I defensively hunker down, hoping she'll leave me alone. Of course not. She approaches the bar and positions herself steps away from me, asking the bartender for kvass. He's incapable of understanding her thick Russian accent, and even if he did, I'm sure he has no idea what the hell kvass is. There's probably no one within a thousand miles of here that serves that bitter, old drink now that Kluge is dead. He used to brew it and gave me a glass on my sixteenth birthday. Kluge, the damn old fool.

"Cola-pop!" Anna angrily snaps at the bartender, then mutters some crude Russian phrases that make me choke on my beer. I quickly try to hide my reaction, but not quickly enough. Anna makes the connection—I understood what she said.

Relentlessly, she begins peppering me with questions. Can't she mind her own damn business? No. That would be easy, and nothing in my life is easy. She's obviously not going to lay off unless I spell it out for her. *Stay the hell away from me.* But when I look up to confront her, the intensity of suffering in her somber brown eyes crushes my determination. Using what little willpower I have left, I manage to spit out, "Talk in English."

Her fury instantly whips through me, then her frustration and then her bitterness. These I can withstand, but when her mood swings to despair and vulnerability, I'm done for. Seeing a tear run down her cheek is my final undoing, the fight to shut her out is no longer winnable.

What the heck? She takes off and chaotically bolts for the door. "I'm sorry," I yell after her, feeling ashamed for upsetting her so. Anna continues stumbling for the exit, and I futilely rack my brain for any other option because speaking fluent Russian in a public place will be a colossal screw-up. No one is to know I can speak Russian, Kluge made that very clear. Yet I can't think of anything else. "Простите," I call out to her.

That stops her like a brick wall. I reach her side, apologize again and find myself trying to explain the unexplainable, that I'm feeling *her* pain. She starts swooning like she's about to faint, so I gently lead her out into the cool night air. We get about a block away before she breaks down and starts crying hysterically. Holding her close, I give what comfort I can with all our combined torments slicing huge gashes into my heart. The viciousness of our joint pain is indescribable.

Eventually, she calms down, tries to push away and then *apologizes to me*? I can't figure her out, I can't figure any of this mess out. Why am I feeling what she's feeling, and what's causing her such incredible pain? She doesn't answer when I ask, instead she turns the question back on me. Normally I wouldn't respond, but nothing about this evening is normal.

After hearing me voice my sorrow, Anna unexpectedly leans in so close that our faces almost touch. The intimacy ignites an irresistible attraction to her. *Remain detached, Joe. No personal relationships.* I should go ... yet my lips touch hers, and my whole world goes up in flames. Everything's gone, except for Anna and me. There are no problems, no internal conflict and no regrets. There's only this incredible feeling. It's amazing!

Then all my wicked, little demons force their way back, spiraling my guilt over Lina out of control. This is a huge mistake on so many levels. Disentangling myself from Anna, I rush her back to the pub entrance, utter a lame quasi-apology and take off walking and walking and walking. Screw the cost, I call a taxi, wanting to get as far away from her as possible. The instant it pulls to the curb, I jump in.

During the entire fifty-minute ride, I wrestle with my guilt because not only did I kiss Anna, I relished the heated contact. I betrayed

Lina, and for a fleeting instant, I even forgot she even existed! Lina's still out there somewhere, and she needs me.

"Sure this is where you want to be, kid?" the driver asks.

"Yes." I hand him the money and step out. The Space of Invisible Souls, Minneapolis's largest homeless encampment, is dark and quiet at this time of night, except for the traffic noise from the road. A mass of makeshift tents fills the small spot of land between the retaining wall and the nearby avenue. Scanning the area, I see a few figures seated under the bridge and head their way. Their body language displays fear and aggression as I approach. Pulling out my cell phone, I bring up a photo of Lina and ask them if they've seen her. After taking a cursory glance, they shake their heads no.

I'm not going to stop looking for you, Lina. I'll search every abandoned house, every dark alleyway, every vacant warehouse, every park and every homeless shelter in all of Minnesota, and then I'll search them all again. I won't stop until I find you.

Chapter 6

In 1935, Albert Einstein, Boris Podolsky and Nathan Rosen discovered a strange phenomenon. Particles that physically interact with each other become permanently connected and dependent on each other's states. These particles then have the ability to instantaneously influence each other, even if they are on opposite sides of the universe.

A different look at the standard explanation of quantum entanglement—Two particles, Anna and Joe, become entangled and are sent to labs of two different physicists named Alice and Bob. When Alice performs a measurement on her particle, Anna, it changes the state of the universe and seals the fate of Bob's particle, Joe. These two now share a common history, and their futures will be forever intertwined.

Tatiana
Late Saturday evening, September 22 to Wednesday, September 26

"That witch Mizuikova started wildly screaming at *me* when she returned from the bathroom and you weren't in the pub anymore. If she wasn't such a bitch, I would've told her that you left with Joe. Not that it mattered, she went poking around and probably found out anyway. What does matter, Anna, is that you should not be getting involved with him." Hannah decrees this with an air of authority as she zooms into our room. "The only thing Carleton cares about is football. His heart is a thumpin' gizzard and he treats women like we're egg-suckin' dawgs, barely even acknowledging we exist. And the few times he does deign to speak to us, it's to say, 'Go to hell.' I wouldn't walk across the street to piss on him if he was on fire." Dramatically placing her hands on her hips, she adds in a loud *humph* to emphasizes her unbidden advice.

Ofiyet'. Hannah must be extremely angry because her colloquial expressions are off the charts. *Hmm.* Can't believe I'm contemplating this, but I might need to learn some of her unique vocabulary because with my current, woefully poor comprehension of Hannah-talk, I'm not even understanding a third of what she says. While that isn't important right now, it could be in the future.

Hannah flits over to her bed, back to her desk, back to her bed, then across the room to me. "Mizuikova intercepted me just when I was getting up to go after you. By the time I got free of her and went outside, you were nowhere to be found. Spill the beans, Anna. What did Carleton say to you? Why did you get in such a tizzy? Where did you go? And was that a Russian word he flung at you?"

I don't reply to any of her litany of questions. Unfortunately, this closed-mouth strategy is becoming ineffective. Instead of discouraging Hannah, it intensifies her interrogation. Evidently, she interpreted my feeble encouragement of camaraderie earlier in the day as the directive she needed to assign herself to the role of caretaker. *Yolki-palki!* Not at all what I intended. To correct her misconception, I'll blatantly shun every one of her inquiries and various overtures of friendship.

I can't be chummy with her, close relationships don't work out for me. I inevitably end up being labeled a ruthless, cold-hearted, self-serving bitch and castigated to the outskirts. That outcome can't happen with Hannah since it's highly likely I may need her assistance again. She's a valuable resource due to her vast network of social contacts and intimate knowledge of the inner workings of Carleton College, information I can't readily get off the internet or through other means. Therefore, it would be foolish to compromise all that by becoming "buddies."

Holding to this solid line of reasoning, I don't utter one word to her for the remainder of the weekend. It takes all the way to Sunday evening until she finally relents and our relationship returns to its previous uncommunicative state.

I, however, am totally incapable of returning to my previous state. Despite Hannah's fierce counsel to forget about Joe, I can't. The residual energy from his kiss continues to thrum wildly through my body. How did this happen? I still have no idea. My reconnaissance mission at the pub to gain more information about him, in order to nix this inexplicable connection between us, somehow went sideways and made the bond even stronger. I should have left immediately when things started going awry, but I didn't, and now I have an additional burdensome problem. When we kissed, I became totally unaware of my surroundings. This irresponsible lapse of awareness could have had Herculean consequences. For like Hercules, I have enemies plotting against me too, and losing sight of that is unacceptable. Until I can determine an effective method of preventing blank-outs like that, I have to keep my distance from Joe. Settling into bed, I burrow deep under the covers and hope I can maintain this firm resolve tomorrow.

The second Joe spots me on his way to chemistry Monday morning, he double-times it to avoid getting in close proximity. It seems he has also concluded that any contact between us is dangerous.

As I enter the classroom, I overhear the instructor talking to him. "Mr. Carleton, a word with you please." Slowing my pace, I strain to catch the rest of their conversation, hoping to understand it. "It has come to my attention that you speak Russian. I would like to reschedule your lab time so that you can attend at eleven a.m. on Thursdays to assist our international exchange student, Anna Alkaeva. Let me know if there's a reason that can't be done."

Well, I certainly caught the gist of that. Not good, not good at all. Taking my seat, I push my back into the chair, bracing for Joe's reaction. His intense hostility hits full force, literally knocking the breath out of me and pummeling any chance I had of concentrating on chemistry. In fact, it booted it into right field. *T'fu.* Yet another foolish American expression. It's not the right field, it's the wrong field. *Pay attention, Tatiana, you can't continue to fall further behind in this class.*

I can't pay attention, not with this wretched, wicked connection. Even though it's unwanted by both of us, we can't break free. What if it expands to encompass the emotions of anyone near me? Right now, it's only Joe's feelings I can't block. What if this happens around Hannah? *Gavno!* I can't have her swirling around in my head. She's an out-of-control, emotionally charged whirling dervish that I would never be able to manage.

Alarm and dread mix into a hideously noxious potion and force their way down my throat, causing me to gag. Swooning, swooning, swooning ... I'm about to fall into the bottomless abyss, to tumble five hundred and twenty-four meters down to the depths of that blasted Siberian Mir diamond mine, the largest excavated hole in the world. Curse that stupid mine and curse its bloody diamonds. Each one of them has given me nothing but trouble.

"Anna, are you paying any attention?" Ms. Mizuikova scolds me.

"No!" With extreme difficulty, I struggle through the remainder of class. All the extra energy it took to accomplish that makes me ravenously hunger, but it's not my lunchtime yet. Craving Corn

Nuts, a current obsession of mine, I swing by the student store and purchase six bags. The sensation of grinding Corn Nuts between my teeth is quite gratifying, and there's the added bonus that the sound drives Ms. Mizuikova mad. The pained expression on her face is so delightful that I've taken to crushing them with my mouth open.

Regrettably, I'll have to forgo crunching down an entire bag in her presence because I require some alone time. Upon reaching my dorm, I firmly instruct her that I'm going to study *by myself* and she need not come back until it's time for my afternoon class.

Incredible, without any argument, she spins around and scurries away. The six bags of Corn Nuts must have influenced her decision making. I'll have to remember that trick for future use.

There's a light bounce in my step when I enter the small cubicle that poses as my dorm room. Plopping down at my desk, I start munching on my snack while aimlessly staring at my chemistry textbook; this solitude isn't for studying, it's for puzzling out the link between Joe and me. I begin by meticulously analyzing every interaction we've had. I'm smack-dab in the middle of Saturday night's events when Hannah bounds into the room. Her visually disarming wardrobe blasts into my field of vision and momentarily confuses me.

How can someone of such small stature, without saying even one syllable, cause such a dramatic disruption in my thought patterns? I blink my eyes and take a swallow of water to get myself back on track.

"Vere did Joe grow up?" I casually ask. It's likely Hannah may have some useful background knowledge on him.

She halts mid step, tilts her head sideways and eyes me intently. Unsure what to make of me speaking to her after days of silence, she hesitates. Then her "chummy" programming reactivates. "That game isn't worth the candles, Anna, he's only going to hurt you."

"Vere did Joe grow up?" I repeat.

"You need to forget about him." She arches her eyebrows and adopts the strict tone of a primary school teacher trying to corral a room full of six-year-olds. "Stay away from him. Joe is a humpback that will be fixed by the grave."

"Vat?"

"Don't you know your own Russian sayings, Anna? It means he's never going to change." She rolls her eyes at me like it's inconceivable I didn't know this.

All right, asking Hannah was a mistake. She's not going to tell me anything useful, so quizzing her further is pointless. My next viable alternative is the internet. However, knowing my father as well as I do, it's highly probable that in addition to tracking my cell phone usage and having Ms. Mizuikova looming over me, he's also had my computer hacked to follow my online activities. Since I don't want him finding out about my heightened interest in Joe, I can't use my laptop.

Fortunately with today's timely turn of events, no Ms. Mizuikova spying on me right now, I can relocate to the library without anyone being the wiser. In short order, I'm settled in a relatively private alcove, logging onto one of the college's computers. Opening Google, I type in *Joseph Carleton, Minnesota,* then cut-and-paste the results into an English-Russian translating website.

Joseph Earl Carleton is an All-American football player for the Carleton Knights. During his freshman year as starting quarterback, Carleton completed the season with 4,805 passing yards, 1,240 rushing yards and 48 touchdowns. The Knights finished with a school high 9-1 record, winning the MIAC title. Mr. Carleton was selected to All-Conference and First-Team All-America, and he was named Sporting News College Football Player of the Year ...

This is totally irrelevant. I click on another site.

Joseph Earl Carleton, golden boy of the Hibbing High School Bluejackets football program, is the number one-rated quarterback in the state and has been nationally recruited to play for Alabama Crimson Tide. During his four years as the Bluejackets' starting varsity quarterback, Carleton set new school records in total passing yards, passing touchdowns, number of passes attempted ...

High school? That must be the same thing as secondary school in Russia. *Blin.* Other than learning he went to school in Hibbing, this information is useless too.

Wait a minute, play for Alabama? Then why is Joe at Carleton College? I scour the internet for more details, but don't get any. Looks like I'll have to figure this out the old-fashioned way, by personally confronting the source, because there's no way I'm asking Ms. Mizuikova or my father to look into it.

Regrettably, that plan entails waiting two whole days since I don't see any way to seek Joe out without Ms. Mizuikova sticking her neck into the middle of it ... unless I lie. No, it's not worth it, not yet anyway.

It takes forever for Wednesday to arrive. Uselessly biding time is not one of my fortes; I'm accustomed to acting on my plans immediately. On the plus side, with almost forty-eight hours of stored up anticipation, I'm primed and ready for a face-off with Joe.

To avoid Ms. Mizuikova's interference, I leave the dorm half an hour early and spend the extra time pacing a ways off from the entrance to the chemistry room. I wait, and I wait, then I wait some more. He doesn't show up for class, and I'm left stranded with a highly irate Ms. Mizuikova, who's demanding to know why I took off so early without notifying her.

There's a slew of obnoxious replies queuing at the tip of my tongue, but I manage to stifle almost all of them. The result is an entire day of brooding which continues into the night.

By Thursday afternoon, when at long last it's finally time for chemistry lab, I storm into the room, stomp over to my assigned table, scowl at the clock and watch the minutes tick off. 10:57, 10:58, 10:59. No Joe. The instructor starts giving instructions and Ms. Mizuikova starts translating. 11:00, 11:01, 11:02. Still no sign of him. 11:03, 11:04, 11:05. Slowly and intangibly, like the wispy tendrils of a jellyfish, strands of Joe's wrath tangle themselves around me and start simultaneously stinging.

Bring it on, Carleton!

Joe enters the room, and the unchecked blast of his resentment crashes into me so hard I think my skull's about to explode. "Stop it!" I scream in Russian. My outburst reaches every nook and cranny, and everyone turns to look at me. No one, however, has any idea why I'm so upset. No one except Joe. Reluctantly, he approaches the lab table with his fury still pounding into me.

"Stop what?" he demands in English.

"Stop targeting your anger at me. I didn't ask for you to be here," I reply in Russian. The intensity of his emotions lessens, but it's still far from a tolerable level. "And if you dare tell me to speak to you in English, I'll deck you."

Joe stands there mutely glaring at me. *Pizdets!* I didn't want to question him under these conditions—Joe combative and confrontational and Ms. Mizuikova hovering next to me—but I'm not waiting for ideal circumstances.

"Why are you here at Carleton?" I whisper as discreetly as possible through the side of my mouth. "Shouldn't you be in Alabama?" Purposely I poke the tiger, hoping his temper will work against him and lower his guard into unwittingly revealing personal information.

"If I had a choice, I wouldn't be here," he bitterly gripes.

No details there, but at least he's talking in Russian. That's encouraging. "You always have a choice," I lamely say, unprepared for the type of answer he gave me. Ironically, my father often tells me this same thing while taking away my choices.

Joe mutters something too softly for me to hear, then picks up a lab packet and starts reading. Conversation ended. That went nowhere. Grabbing my packet, I distractedly glance at it and ask Ms. Mizuikova to reread the instructions. She doesn't say a thing. What's wrong with her now?

Looking over, I see her intently staring at Joe. *T'fu.* I shouldn't have been so reckless. My hasty decision to charge ahead only ended up giving her more ammunition to pass on to my father. Although, after that incident at the pub Saturday night, it's a sure bet my father has already started digging deeper into Joe's background, and unlike me, he's probably discovered the reason Joe isn't attending college in Alabama.

Ms. Mizuikova menacingly flickers her eyes between Joe and me a few times. Tightly pursing her thin lips, she begins reading the lab material in a clipped tone. "Finding the Molarity of an NaOH solution. Step one, clean the burette with water. Step two, rinse the burette three times with the HCL solution of 0.1 molarity. Step three, fill the burette with the HCL solution and remove any bubbles. Step four ..."

My concentration wavers, slips off the page and spreads over to Joe. His eyes aren't moving. He's only pretending to read. What is he really thinking about right now?

"Step eleven, using the formulas, calculate the molarity of the NaOH," Ms. Mizuikova concludes.

Well, I missed most of that.

Joe continues "studying" the packet while my other lab partner and I gather the necessary equipment and solutions. Once everything's arrayed on the table, Trisha jumps right in, cleaning the burette and rinsing it with the HCL solution. After completing a few more steps, she nods in my direction. "Add the indicator, Anna."

Huh? I'm not sure how to proceed. I don't remember much past the third step. Trisha impatiently taps out a few bass drum thumps with her foot, then grabs the burette.

"No," Ms. Mizuikova firmly rebuffs her. "Anna needs tu partizipate. Give her de burette."

Trisha obediently hands it over, and I add some drops to the beaker while she stirs. Magically, the mystery solution turns pink.

"Okay, add the HCL solution, Anna," Trisha instructs. Getting the other solution, I cautiously start adding drops, but with Joe still fuming just a few meters away, I'm having extreme difficulty focusing. My thoughts quickly become muddled.

"Anna, stop!" Trisha shouts. The mystery NaOH solution is completely clear. Trisha keeps stirring and stirring, but nothing happens.

"The mixture's still clear because you added too much HCL. The excess bonding is preventing the indicator from turning back to pink. You need to start over," Joe lectures me in Russian.

"No. *You* need to start over," I correct him. "Go get another beaker of the NaOH solution."

"Wasn't my mistake," he smugly clips back, switching to English. As I wind up my fist to deck him, he grabs ahold of my forearm. The contact triggers an intense heat to sear through the sleeve of my shirt into my skin, and I pull back, spooked by the reaction.

"Get another beaker," I yell.

"Wasn't my mistake," he jabs another barb at me in English.

He's right, I made the mistake; I totally miscalculated how this "interrogation" of him would turn out. Needing to get off the war path, because here is not the place for a battle, I march off to regroup.

Chapter 7

Joe, age nine

"What the hell's wrong with you?"

I've recently figured out that when my father asks this question, he doesn't really want me to answer.

"Do you have any idea how tired I am of leaving work to talk with the principal about you?"

Again, another question I shouldn't answer. I keep staring out the car window.

"Pastor Gordon suggested you start helping an elderly German man who moved here recently. He doesn't speak much English and lives just outside of town. Starting next Monday, the bus will drop you off after school at Mr. Rudolf Kluge's house."

Just freaking great. Since my father's German and I speak German, I get punished with this. Instead of hanging out after school at the clearing near the bus stop to play football, I'm going to be stuck with a grumpy old man barking orders at me. One more way my life gets messed up.

Joe
Thursday, September 27, 2012

The moment Anna takes off to get another beaker of the NaOH solution, her prison guard of a translator, Ms. Mizuikova, levels her steel-gray eyes on me. Disdain is branded across her harsh features and splits outward from her deep crow's-feet. The old battle-ax's entire face is twisted up, and her tightly coiled black hair writhes with contempt. Snakes seem to be slithering and hissing out of the bun on top of her head.

"I know your type," she sneers in Russian. "You think the world should be gifted to you, that you're entitled to bypass life's hardships, because Americans are obsessed with their sports and give special allowances to their athletes. Well, you're nothing but a dvoyechnik, and that's all you'll ever be."

Don't let her push your buttons, Joe. If A is success in life, then A=X+Y+Z, where X is work, Y is play and Z is keeping your mouth shut. Fixedly, I keep chanting Albert Einstein's advice, to keep my mouth shut.

"Anna's not a dvoyechnik like you. She's an excellent student, and it's only because of your juvenile interference that she is not succeeding in this class." Her beady eyes drill into me.

Well, that's all I can take. "Piss off and go bare your fangs at someone else."

Her face contorts with a wicked grin, pleased that I went after her baited hook. Stretching her neck a little higher, she strikes again. "When Anna disappeared from the pub Saturday night, I questioned the bartender. He told me she started crying after talking with you. Since he doesn't speak Russian, he didn't know what you said. So educate me, what is it you tell women to make them cry?"

"Go to hell," I truthfully answer.

A barely perceptible tic appears near her eye, followed by her forked tongue flicking in and out. "You'll be there long before I will. You are a wild, dangerous deviant who should be locked up. I don't understand why this justice system allows you to roam free after you were arrested for assault last year."

Shit! Someone's been very effectively prying into my past. No sense arguing with her and clarifying I was never charged. Instead, I mouth "fuck off" at her.

This time she jerks back almost a full foot. "I will not stand by and let you torment Anna. Your selfish, hurtful, destructive influence is making my job impossible. End your childish harassment, or I will!"

Torment Anna? All I've been doing recently is staying the heck away from her. Yet, I've been so wrapped up in my own problems, I haven't considered how our strange connection might be affecting her. As I begin analyzing it, the Gorgon Mizuikova hisses more threats at me. Venom is dripping off each of her words. *Don't do it. Don't take the bait again, Joe.* Grabbing my backpack, I shove off before I do something stupid.

When I get back to my room, Lance looks up. "Aren't you supposed to be in chemistry lab right now?" He's wearing his black-rimmed glasses, the ones that give him a scholarly appearance. Weirdly, his closely cut, wiry hair looks even shorter than it did two days ago.

"Did you get *another* haircut?" I ask to divert him.

"Why are you back so early, Joe?" He scowls at me while the sweat speckling his dark skin starts dripping down his face. Picking up a neatly folded hand towel, he wipes it off, then refolds the towel and lays it back on his desk. Last night, the heating system kicked in and wouldn't quit; our room quickly became a sauna, even with the windows open.

"Thanks for your concern, Sir Lancelot. Inform the Round Table all is well." Crossing one hand over my chest and the other across my back, I formally bow allegiance.

He's not amused. "Stop calling me that."

"The valiant Carleton knight doesn't like being compared to the legendary King Arthur's knight?" I feign mock astonishment. I really do consider Sir Lancelot a fitting nickname for Lance. Carleton College's official mascot was a knight until the 1970s, and the school's varsity sports teams are still called the Knights.

"No, I don't like it," Lance states. "Lancelot committed adultery with Arthur's wife and was ultimately responsible for destroying the Round Table. Therefore, I don't like the connotations associated with it."

He does have a compelling point. Unlike King Arthur's Lancelot, this Sir Lance is loyal to a fault. He's also conscientious, courteous and reserved, all qualities that seem to have eluded me.

"What's going on with you now, Joe? How come you aren't in chemistry lab?" Even though Lance is irritated with me, his face shows concern. Besides being the lead running back for our football team, he's also my roommate and closest companion. Unlucky for him, the coaching staff seems to depend on Lance to help keep me reined in.

"Finished early," I say. Taking three books out of my backpack, I drop them on my desk and they land with a loud, satisfying *thunk*. I picked them up at the library earlier, selecting each for its different perspective on the Minnesota Viking's greatest players. Fran Tarkenton and Alan Page are my current picks.

"Let it go," Lance counsels me. "You already fulfilled your once-a-week quota of being a royal ass on Saturday night.

I can't let it go! I'm cursed to continually remember every single second of that damned day, September twenty-second, 2009, the annual passage of Lina vanishing into thin air. For months afterwards, I recklessly charged around searching for her, but I couldn't find even a trace of her whereabouts. To escape the brutal guilt, I started doing every extreme sport I could think of with a kamikaze attitude. It didn't help. Nothing helps.

Lance scrutinizes me closer. "Something's different. It's that Russian girl, isn't it?" I glance over at his laptop instead of answering. "Project Zooniverse," he readily explains. "It's a citizen web portal where volunteers participate in scientific research. I'm one of the citizen scientists helping with the Planet Hunters project. We're identifying planets outside our solar system from the light curves of the stars recorded by the Kepler telescope. More specifically, we're looking for Earth-size planets orbiting a star where life might exist." He's getting all spooled up to describe in exacting detail another of his CitSci missions.

Usually, I put a quick stop to that, but today I don't interrupt, glad the subject of conversation is off of me and my problems. On and on Lance rambles. I change clothes and get ready to go running, needing to work off some of my copious frustration. When I bend down to tie the laces on my new running shoes, I cringe in disgust, not yet accustomed to their obnoxious lime-green color. The sparkly silver reflective material along the sides and near the heels is only slightly less in your face. My sister Beth bought these for my birthday and they're her preference in colors, not mine. But I don't have the money to buy another pair any time soon.

"After hearing Natalie Batalha discuss Project Zooniverse on Minnesota Public Radio, I joined the project," Lance continues his dissertation, "She's a member of the Kepler team and the one responsible for selecting the roughly 150,000 stars in the portion of the Milky Way we're examining."

Alright, I've heard enough. "Later, Lance."

"Hold up, Joe." Not wanting to lose his only audience, he bolts out of his chair, almost knocking it over in the process. "Natalie Batalha has this fascinating view about science. Connectedness."

I make a break for the door, but he anticipates my move and skillfully blocks the way.

"We're not *apart* from the universe, we're *part* of it. Think about it. Curiosity rover climbing the ridges on Mars, Cassini orbiter tasting the dust of Saturn, Voyager spacecraft seeing the entire solar system and now Hubble telescope helping us view other galaxies. By reaching out to the universe, we're reconnecting life on Earth to life beyond Earth!" he jubilantly exclaims.

"Great. See ya." Feigning to the left, I try to dart around him to the right, but he cuts me off again.

"With this Planet Hunters project, you can be part of discovering the galaxy too." There's a hopeful pitch to his voice. Lance doesn't concede defeat easily, I'll give him that, but he's not going to talk me into joining his present undertaking. It's been a full year since he duped me into helping him with Project Squirrel, a different CitSci mission to aid scientists in better understanding tree squirrel ecology. My annoyance from being roped into that project hasn't faded yet. After training me as a squirrel monitor, Lance deployed me to different neighborhoods to identify and count tree squirrels. I naively assumed it would be a day or two evolution in the local area, not a long-term commitment to report on the majority of southern Minnesota. I lasted four days before I bailed. No way am I getting cornered into another of his pet projects.

"That's an amazing transition, connecting your thoughts like that, Lance. I'm impressed, and I'm also chuffed." I wait for his reaction. We have a long-running competition of trying to stump each other with relatively obscure words. *Chuffed* is a British term that since the mid-1900s has been used to mean both *pleased* and *displeased*.

The puzzled look on his face morphs into a *chuffed* expression. Being one-upped, he grudgingly steps aside, and I head out for a long run. The miles pass and my muscles gradually begin to unwind, but my brain won't budge. It keeps rehashing the same question. How is Anna messing with my head? For the past year, I've been able to keep myself mostly in check. Now all her clashing controversies are screwing with that. A great *Star Trek* line gives me some comic relief, although in Anna's case it has to be slightly altered. *Woman, a mass of conflicting emotions.*

I pick up the pace a bit more and it helps distract me. By the time I'm back on the main part of campus, I'm almost relaxed ... then Anna appears, and this bullshit connection that's hardwired into our systems fully engages, spewing her unadulterated pain into me. *Damn Lance and his connectivity!*

Swiftly changing directions, I take a different route back to the dorm, yet the bitter truth doggedly runs after me. Fighting this obscure link is only making it worse for both of us; it's ripping me apart and it's ripping Anna apart.

"Still all *chuffed* about that Russian girl?" Lance bags me as I enter the room.

"Piss off, Lance."

"I know it's way outside your normal course of action, Joe, but you could just talk to her." He throws this bit of wisdom over his shoulder as he leaves for his philosophy class.

Not happening. Talking is not a good option, not when it comes to women and especially not with women attracted to me. *Talk* is a cursed four-letter word: Tedious, Aggravating, Laborious and Keenly difficult. All around, a bad idea.

I grab one of the footballs I keep in my room and start doing gripping drills. After ten reps without dropping it, I sit down and close my eyes to give myself more of a challenge. I finish two more reps before my grip slips and I have to open my eyes to find the football.

Crap. Lina's cousin is standing in the doorway, solemnly staring at me. Her long, black hair, which hangs just past her shoulders, sharply contrasts with the faded pink colors she's wearing. Her face is paler than I remember, causing her dark eyes to look even darker. After cautiously glancing around, she tentatively eases into the chair by my desk.

"I came to say goodbye, Joe. There's nothing more in Hibbing for me. I'm leaving for Florida to get a fresh start." She expectantly pauses, uncomfortably shifting in her seat while waiting for me to say something. When I don't reply, she reluctantly continues. "My cousin Russ started classes at Minnesota State University in Moorhead. He'll never forget what you did for him, getting kicked out of school and all."

For the first time since she entered my room, Ginny looks directly into my eyes. I know exactly what she sees there, because I can't hide it, the guilt that plagues me every day.

"Ellie's never coming back," she announces with an air of certainty.

"It's Lina, not Ellie," I sharply state. I hate the name Ellie.

"You're the only one who calls her that." Ginny nervously rubs her fingers together. "She had no choice, Joe. With no money, no high school diploma, no skills and no one to turn to, what else could she have done?" Her sullen statement falls heavy between us, and we sit there, each absorbed in our own penance, wrestling with our own misgivings.

Ginny isn't just Lina's cousin, she's her best friend. Ginny was the only one Lina reached out to, and she didn't do a damn thing to help. I can't stop blaming her. "Lina could have come to me," I argue.

"She was afraid of what you would do, Joe."

Another rock piled onto my mountain of guilt. "Has she ever contacted you?"

"No." Ginny folds her head to the floor before hesitantly conceding, "I know she left for the Twin Cities. She was hoping to find some sort of work there, but since she was only seventeen and didn't want to be discovered and sent back home ..."

Her unfinished sentence sends my mind along feeds I don't want it to follow, the brutal statistics of teenage runaways: sixty percent become prostitutes for a time to survive, the average prostitute gets physically attacked once a month, and their death rate is 5.9 times that of all teenagers. "I don't care what she's done," I plead, trying to shove away these hideous facts.

"Maybe you don't, but Ellie does, and it will always be there between you. By staying away, she's saving the love you both still have for each other."

"That's bullshit, and you know it, Ginny." I clench my fists, trying to calm down.

"You have to let her go," she whispers.

"I can't let her go!" Why can't anyone understand that?

Ginny shrinks back in the chair, then quickly stands up to leave. Halfway to the door, she slows. By the slight bobbing of her shoulders, I can tell she's starting to cry. "I have to live with it every day of my life, too, you know." She bolts into the hallway, barreling right into Lance. After awkwardly skirting around him, she races down the stairs.

"Another happy patron of the Joe Fan Club, I see. Who is she?" Lance gives me a stern look.

"Go bother someone else, Lance." *You don't always have to be the knight in shining armor.*

"Guess you're a little *chuffed*. I'll just drop off my stuff and leave you with all your admirers then. Have fun."

Fun. Sure thing. Leaning back, I rack my memory over and over, continually spinning on the same track, like a gerbil on an exercise wheel. Ginny's logic doesn't work for me, I can't let Lina go. My brain doesn't work that way; it's not a switch I can turn on and off. And there's always that remote hope of finding Lina that wickedly taunts my waking hours and haunts my dreams. It never goes away—except for that one brief moment Saturday night when I kissed Anna, and I can't come to terms with that.

Yet avoiding, ignoring or blaming her isn't resolving anything, it's only making things worse. For both our sakes, I have to suck it up and deal with it.

Chapter 8

To know more, one must feel less. But to know the soul is through the heart, not the mind. Thought cannot unravel the mysteries of creation because the mind is an instrument, a machine ... It is the soul that is the true medium for attaining the highest knowledge.

Fyodor Dostoevsky
Based on a letter to his brother in 1838

Tatiana, age eight

I don't understand. Why is Father so upset with me this time?

"Tatiana, that type of behavior is totally unacceptable. Do you understand?"

What type of behavior? What did I do wrong?

Tatiana, age twelve

I still don't understand.

"Tatiana, what you did is disgraceful. Why can't you act properly?"

Years of watching people interact, trying to figure out when to laugh, when to act sad, when to pretend to be happy, and I still don't get it. I feel like a puppet who is only playing at being human.

Why can't my soul sing?

CHAPTER 8

Tatiana
Friday, September 28

I don't *understand*! After Joe's bitter participation, or lack of it yesterday during chemistry lab, I should be furious with him. Instead, there's a strange dichotomy rippling between us, a coupling of resentment and reconciliation, aversion and attraction.

When class ends, we glance at each other and a meaningful look passes between us, yet we're both reluctant to approach one another. Joe makes the first move and apprehensively works his way upstream against the current of departing students.

"Can we talk privately?" He jerks his head in the direction of Ms. Mizuikova.

Well, he's back to speaking to me in Russian. This sudden reversal of attitude is confusing. For the past week, he's either ignored me or been openly hostile. Therefore, I'm not sure I want to listen to whatever it is he's going to say. "We can talk right here," I reply.

Leaning in close and lowering his head next to mine, Joe whispers in my ear, "We are the ones we've been waiting for."

Instantaneously, Ms. Mizuikova thrusts herself into action, pushing Joe away from me and menacingly threatening him. "If you dare harass Anna any further, I will personally ensure you spend the rest of your life regretting that decision."

"Can you call off Medusa?" Joe snaps, his face flushing with anger.

His reference to Medusa sparks an image of Joe as King Arthur, clad in shiny armor, battling the monstrous Gorgon, Mizuikova. After all the incredible stress and tension of this past week, the bizarrely comical confrontation makes me giddy, and a snort escapes me.

"Get away from Anna, immediately!" Ms. Mizuikova demands as waves of animosity stream across her features.

I instantly sober up, realizing this is a fight that Ms. Mizuikova, backed up by my father, would win. "I'll handle this matter myself," I direct her.

"I am not going to—"

"Leave, now!" I firmly instruct.

She starts visibly shaking with rage. After glaring a few more daggers at Joe and me, she spins around and struts out of the classroom to wait just outside the door.

Really, Father, couldn't you have hired someone with a little more self-control? Refocusing on Joe, I attempt to determine his intentions. His Siberian blue eyes are cold and brooding, yet also strangely inviting. "Why did you tell me we're the ones we've been waiting for?"

"It's a Hopi expression. I interpret it to mean that the solution to a problem lies with the people involved. Our fate is in our hands."

The expression bumps at me for attention like a teenage girl poking at her beau. Unsettlingly, the concept of *our fate* expands to encompass more than just Joe's and mine. Is what's transpiring between us linked to the world on a greater scale?

"Anna, let me help you with chemistry."

He's got to be joking. The suggestion is so absurd I dismiss it without consideration. "Too much more of your help and I'll be ensured of failing this class."

"Can we agree on a truce? I feel somewhat responsible for your difficulties. Let me tutor you."

The pathetic look on his face almost makes me consider it. "I don't trust you. Why aren't you playing football for the University of Alabama?"

An uncomfortable silence follows, and Joe becomes extremely irritated, much more than I think he should be. Lining up his foot with a forgotten pen on the floor, he forcefully kicks it, sending the implement ricocheting off numerous chairs. Joe follows its progress until it disappears from sight, then a resigned expression settles over him. "I was expelled from high school and lost my scholarship."

"Then how did you get into Carleton?" His explanation doesn't make any sense.

"My football coach." He states this as if this clears up everything.

"Carleton is ranked the eighth-best liberal arts college in the United States and has a history of enrolling students who receive National Merit Scholarships. It's a highly selective institution and doesn't offer sports scholarships." I did thorough research before I selected this school.

He seems oddly amused by comments. "Did your pet Gorgon, the one with serpents writhing on her head, inform you of that?" Jokingly, he adds in a few hisses at the end.

"No, I was able to accrue this information all on my own. I am literate, you know." The hint of a lopsided grin forms on his face, and I scowl at him to show my annoyance. *Pizdets!* Now his smirk stretches from ear to ear.

"Would you prefer I call her Lady Macbeth or perhaps Queen Mary I? Although, I should probably find a Russian moniker to pin on her. What do you think, Anna?"

Enough of this useless banter, that's what I think. "Just tell me how you got into Carleton."

"All right. I took the GED and received my high school equivalency. Then I retook the ACT and got scores that qualified me. My high school football coach is friends with the admissions officer and explained my situation to him. Does that satisfy you?"

Huh. All I have to do is repeatedly ask him direct questions, and he'll tell me what I want to know. Could it really be that easy? I doubt it. "So, since your last name is Carleton and you're attending Carleton College, you have connections, some strings were pulled and here you are." I fill in the missing details. "Is that correct?"

"Why do you need to know of all this?" His eyes are narrowing with a wary suspicion.

I hit a nerve. Or did I? Maybe this is what he wants me to believe, and his reactions are a ruse to cover up something else he wishes to keep hidden. Possibly he's—*Stop, Tatiana! You can't remain clearheaded enough to succeed in this contest of wits if you let yourself get mired down second guessing his shifting emotions.* I'll just confront him head-on again.

"Tell me exactly how you got into this college. The entire process," I demand.

"I already told you, Anna. Now you answer me. Why do you need to know?"

What should I do? I've reached a critical tipping point. Do I tell him my cover story, or say nothing? I certainly can't reveal the truth. "What answer do you want?" I volley back.

"All of them," he challenges.

It's my turn to carefully analyze the dangers of revealing information. Yet before I have a chance to evaluate my options and their associated risks, an irrepressible urge to confide in him overrides my caution. Recklessly, I blurt out, "Because the few facts I've been able to gather about you don't make sense, because I'm inquisitive and you're a conundrum to me, because my roommate says you're no good, but I draw my own conclusions, and because I have an extremely overprotective father who meddles in my affairs." My rush to get all this out leaves me with no reserve air to send the last thought through my vocal chords. Hauling in a deep breath, I complete the comprehensive list with the most compelling reason. "It's critical I figure out what's happening between us because it would be a *very* bad thing if my father knows more about you than I do."

With each of my remarks, Joe's emotions shifted, which then swayed my intentions, causing me to divulge far more than I should have. This blasted connection that fetters us together is significantly affecting my decision-making.

Zzzap! A deeply repressed memory arises and spreads itself over the inner surface of my corneas, blocking out external stimuli.

I'm sitting on the edge of Mama's bed, staring down at a book I found in a remote corner of her closet. The pages are well worn, and the dust jacket is torn. As I place the book in my lap, it opens to a page she dog-eared. A disturbing photo of the destroyed Chernobyl Nuclear Power Plant blares up, immersing me in its desolation. Foreboding fragments of horror drift through my mind like fog gliding across low-lying marshlands. A mixture of somber sounds joins the fray, echoing the trapped agony trapped.

"Tatiana, no!" Mama's scream startles me, and I look up. She takes a few steps toward me, then collapses to the floor and grasps her head. Instantly, all my senses are violently submerged in her soul-searing grief. The grief of hundreds, yet the grief of one.

Mama regains control, and I'm released. Confused and disoriented, I run into her comforting arms. Holding me tight, she whispers, "My precious little girl. Pray you are spared this."

Spared what? She never told me; she died with her secrets. Why did I just remember that in such vivid detail? I strain to discover any correlation between the powerful memory and my unintended revelations to Joe. Only vague correlations emerge. Violent emotions, hidden secrets and unpredictable consequences. My conclusion, be careful, be very, very careful. I need—

"Carleton's football coach also talked to the admissions officer," Joe tacks on another minor detail.

"Carleton doesn't offer sports scholarships," I repeat. He's back to giving answers that don't answer anything.

"It didn't happen that way, Anna."

"Then what way did it happen?"

"I spent a year working on my father's farm first."

Another non sequitur. If only I had a truth serum, it would make this so much easier. "That doesn't explain how you got admitted to Carleton. And why were you expelled from secondary school? I need the whole truth, Joe, not some vague conglomeration of fragmented facts."

He's getting irritated with me again. Good, because he's sure causing me a hefty dose of frustration. *Whoa!* Now his irritation is getting infused with ... *pity*? This mixed-up feedback isn't congruous. Something doesn't fit. Even for me, whose emotional maturity level is barely above a toddler's, the parts aren't adding up to the whole, or maybe it's that the whole is greater than the sum of its parts. Either way, I'm missing something. Tossing together what I know so far, I try to make sense of it. Possibly, Joe is one of my grandfather's enemies, or he could be an agent hired by the Russian regime. No, I'm still alive, so that isn't it. He could be an opportunist looking to capture me for ransom. I'll keep that on the list. *Hold on! Could it be?* Did my father hire him as an extra set of eyes to keep watch on me? I wouldn't put it past him, and it does explain why Joe speaks fluent Russian. Also, it clears up why my father hasn't made more of an effort to keep me away from him. Everything makes sense except ... it doesn't feel right.

"Are you going to accept my offer of tutoring or not, Anna?"

I don't know. Who are you, Joe?

Chapter 9

The heavenly motions ... are nothing
but a continuous song for several voices,
perceived not by the ear but by the intellect,
a figured music which sets landmarks
in the immeasurable flow of time.
Johannes Kepler

Johannes Kepler was a mathematician, astronomer and astrologer whose visionary work during the scientific revolution of the seventeenth century transformed our view of the universe.

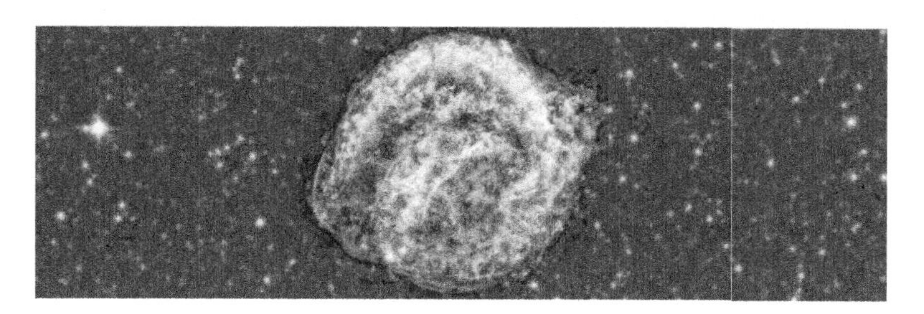

Kepler's Supernova, *Image by NASA/ESA/JHU/R. Sankrit and W. Black*

Tatiana
Saturday, September 29

As we amble back from the library after our first study session, I'm actually experiencing awe. My feelings or Joe's? I'm not sure. Drawing in a breath of cool September air, I gaze upward and soak in the spectacular splendor above us. Flickering and shimmering in all its ethereal glory is the aurora borealis. Tonight, its fluorescent arcs of vibrant greens are especially brilliant.

As I bring my scan earthward and peek over at Joe, I don't perceive any discord. For the first time since I've known him, he's not radiating tension and conflict. Inexplicably, I want to talk with him, not interrogate him, which was my real reason for excepting his offer of tutoring.

Hmm. I wonder what it would be like to simply chat about the cosmos. It's a completely foreign concept for me, having a conversation without a clear purpose. It's also something I should repress for my own safety since who knows what this strange connection might make me spurt out next. Yet, just a short, cursory discussion about innocuous matters shouldn't cause any harm, right?

"When I was very young, I used to look at the aurora borealis and think about living there." Tilting my head skyward, I'm that little girl again and I start softly singing. *"There is a world beyond the stars, I hear it singing in my sleep, there all the troubles of this life—"* I abruptly stop.

My singing is generating an intense melancholy feeling which I briefly mistake as my own suppressed emotion. Then I realize, it's Joe's. I have no clue how to respond, so I revert back to discussing astronomy. "Over the years, I've stared into every image of the universe I could find. One of my favorites is the Butterfly Nebula. It's magical aura transports me thousands of light-years away, placing me inside the nebula itself, to share its last moments of life. How regrettable that such grandeur is the result of the death throes of a dying star." A strange feeling of wistfulness tingles through me.

"Too hot for my taste." Joe licks his finger, touches the skin of my cheek and makes a sizzling sound. A sense of enthusiasm builds within me; I'm excited by the intoxication of what to me is intimate conversation. And even though Joe's spartan comment isn't a ringing endorsement, I take it as a sign of encouragement.

"I'll grant you that point, Joe. At 222,204 degrees Celsius, it is one of the hottest places in the galaxy. Yet as incredible as the Butterfly Nebula is, it doesn't make the top of my list. Kepler's Supernova holds that esteemed position. Its image practically pulses off the screen and then back in again, almost like a heartbeat. It's as if I'm simultaneously perceiving both the beginning and the end. Do you know that Kepler believed the Earth has a soul?"

"You just switched subjects from astronomy to astrology, Anna."

"I'm well aware of that, but it's immaterial." I'm not getting sidetracked by arguing semantics. "Do you believe the Earth has a soul, Joe?"

"You can't just dismiss that distinction as immaterial," he firmly states.

"Sure, I can. Watch me. In his book *Harmonices Mundi,* Kepler wrote—" Suddenly, I'm tongue-tied. Beside a garbage can, lined up like abandoned toy soldiers, is a long line of what appears to be squashed, lopsided, black-and-white skunks rustling in the breeze. "What are those?"

"A huddle of blow-up "Oscar" penguins that seem to have met their demise," Joe explains. "At Carleton, the emperor penguin is special because—"

"Never mind. We're talking about Kepler." No way am I deviating from discussing my favorite astronomer to chatter about inflatable penguins. "Kepler wrote that the movements of the heavenly bodies are an everlasting composition of independent melodies, and the harmonies they create interact with the human soul." My heartbeat is racing, and the back of my neck tingles as I recite this.

"He didn't intend it to be interpreted that way, Anna. Kepler's music of the spheres is a series of mathematical formulas measuring the angular velocities of the planets from the sun. Kepler himself stated that no sounds are given forth." Joe bluntly refutes my assertion.

Those poor, deflated penguins. Did someone like Joe crush their dreams and whoosh all the spirit out of them? Well, I'm not going to end up like that. I've read quite a lot about Kepler, probably more than anyone else attending this college, and I'm fairly certain I'm more of an expert on him than Joe is.

"That is not true. Kepler wrote out musical notations for each planet's movements. In addition, he firmly indicated that celestial music only becomes complete when heavenly tones interact with human souls," I confidently argue back.

"Kepler is not someone to be fantasizing about, Anna. His father deserted his family when he was five, his mother was tried for witchcraft, he contracted smallpox as a child which left him with limited vision and disabled hands, his first wife died from Hungarian spotted fever and only five of his eleven children survived into adulthood."

"All the more reason he's so amazing. Tell me, Joe, what do you think about the link between music and the soul?" I pointedly ask, not allowing him the chance to wiggle his way off of this topic.

"Tell me, Anna, what do you think about Kepler's statement that the darkness of the night sky directly conflicts with the idea of an infinite universe filled with stars?" he counters.

"That has absolutely nothing to do with what I asked!" What's up with him now? Why is he avoiding *this* question? "Do you think music interacts with the human soul? And don't answer with another question."

A brooding mournfulness pours out of him, passes through me, then stretches deep into the night. As I await his reply, strolling along in this unfamiliar, unsettled state of being, my thoughts drift upward, pinging among the nebulae and supernovas. I'm far away in my travels when Joe jolts me back to Terra firma.

"It took him decades, Anna."

"What took decades?" Part of me is still drifting among the cosmos. I'm not sure what he's referring to.

"Discovering the laws of planetary motion. Kepler started on them in 1595 when he discovered that nesting each of the five platonic solids within one another resulted in six layers corresponding to our then-known six planets. Mercury, Venus, Earth, Mars, Jupiter and Saturn. Amazingly, that had nothing to do with the laws of planetary motion he later developed. It wasn't until 1621 that he finished his third law of planetary motion."

Yes, Joe, and those three laws of planetary motion helped him finally solve the Martian problem—the Red Planet sometimes appears to go backwards across the sky. Kepler deduced that since planets move on elliptical paths at different speeds and the time to orbit the sun is based on that, the whole thing is just an optical illusion. But do I say any of this? No, instead, I stand there locked up with my mouth tight-lipped.

"How do you know so much about Kepler?" I end up asking, the words coming out as more of an accusation than a question. It's quite a convenient coincidence, a little too convenient to be random. Someone must have filled him in on my interest of astronomy, and more specifically, Kepler. My father, perhaps?

"Despite popular belief, I am literate," he says with a grin on his face.

Touché. An excellent use of my comeback yesterday, Joe. The emotional pendulum swings back the other way. This verbal sparring is quite enticing, too enticing for me to cut it off just yet. A wee bit more should be all right.

"There's a NASA project named after him. It's called the Kepler Mission and is listed on the web portal of Project Zooniverse. The goal is to find Earth-size worlds where life might exist."

"So I've heard," he wryly mumbles.

Another bout of uneasy discomfort hangs over us as we approach my dorm. Eight meters from the entrance, Joe takes ahold of my hand and halts our progress. "The diversity of the phenomena of nature is so vast and the treasures hidden in the heavens are so rich precisely in order that the human mind shall never be lacking in fresh nourishment."

The entrancing timbre of his voice captivates me, and I get caught up in each of his words. Standing there stock-still, I delightedly absorb their essence. In fact, I'm so mesmerized that when I attempt a cognizant reply, my mouth is left gaping open like I'm a baby bird.

Joe quickly mumbles goodbye and hurries off. Apparently, my reaction spooked him; it spooked me too. I race up the two flights of stairs to my room and rush inside. Flipping open my laptop, I google *Johannes Kepler famous quotes* and click on a link. Just what I thought. Joe's, or more aptly Kepler's, exact words stare up at me. Why would he have memorized this?

Oh, Kepler, this starry night. Twisting, turning, swirling, whirling. Oh, Kepler, this starry night, eerily, it's oh-so-right. The lyrics I interpret differently than their composer intended, but oh do they fit.

Chapter 10

Joe, age nine

I don't want to be dropped off at Kluge's house after school anymore!

Pretending I don't understand German hasn't worked. Maybe if I get him mad enough at me, he'll kick me out for good, and I won't have to come back. Using my pocketknife, I cut holes in the bottoms of all his garbage bags, then I cut holes in the toes of his socks and finally I cut holes in his dish towels.

The next day, when I get off the school bus and go into his house, I can tell he's pissed at me, really pissed, but he doesn't say a thing about what I did. What's it going to take? There must be something that will tick him off enough to throw me out.

When he orders me to bring his tea, I purposely spill honey all over his desk and onto his papers. That does it. He gets super mad, I mean, really, really, really mad. Grabbing me by my shirt, he slams me against the wall and yells into my face, "Yeban'ko maloletnee, sookin syn, Svoloch."

What language is that? It sure isn't German.

Joe
Saturday, September 29

Walking back from the library with Anna after our first study session, we're both actually relaxed. As she gazes up at the sky, I look more closely at her. A breeze is blowing back her deep-black hair, which reaches just below her neck. Her olive-colored skin is almost glowing, and the slight sheen enhances her thick eyebrows and full eyelashes. As we pass under a light, I realize she's not wearing any makeup. For a woman, Anna's relatively tall, only about four inches shorter than me. She has an athletic build and carries herself with a strong sense of self-confidence, but she's weighed down by her heavy emotional burdens.

Therefore I'm surprised, in a good way for a change, when she becomes upbeat and enthusiastic as we start discussing astronomy. Unfortunately, she soon twists the conversation toward music and the soul. *I can't go there, Anna.* For self-preservation, I only half listen to what she's saying, enough to make coherent responses.

"What do you think about the link between music and the soul, Joe?" She keeps pressing me on this.

Music and the soul. The words automatically bring up Lina's face. There's a smile on her lips as she takes a sip from the coffee mug I bought her for her sixteenth birthday. Painted on the side of the mug is an image of Mozart surrounded by music notes shaped into a heart. Underneath is the phrase, *Music is Love.*

Don't go down that rabbit hole, Joe. Think about facts, unemotional, impassive facts. Okay. The Russian Mir diamond mine is one of the largest man-made holes in the world. The De Beers diamond company tried for decades to discover its mysteries. How could the mine produce so many diamonds for so long? And why were all of the Mir gem diamonds, the Silver Bears, the exact same size and shape?

Those secrets are still hidden ... just like the ones between Anna and me. I envision us teetering at opposite edges of the Mir crater, helplessly staring at each other across the expanse as we're pulled down into its dangerous depths. What the heck? Why did I imagine that? What sort of danger is Anna in?

Thankfully, we reach her dorm before more weird things can happen. Reaching for her hand, I stop a ways from the entrance, where there's a clear view of the stars. I want to say something that will leave her upbeat, and I search my memory for a suitable Kepler quote. As I'm reciting it, she stands transfixed. Hopefully, that's a positive sign. Her temporary paralysis works to my advantage, and before she has a chance to question me more about music and the soul, I'm out of there.

On the walk back to my room, I think over how tonight went. Overall, it went well. We covered two chapters of chemistry, Anna didn't have any emotional blowups and we had a somewhat normal conversation. If only she hadn't started singing and philosophizing about music and the soul. That's something I can't dwell on because Lina loves music. She had a part in every school musical and practically lived in the choir room. Music is *her* world. Lina's father claimed that even before she could walk, her mother was already teaching her how to play the violin, and at age thirteen, she started playing with the Mesabi Symphony Orchestra. Music is Lina, and Lina is music. For me, the two are inseparable.

Don't obsess over Lina. Don't totally mess your life up again, Joe! Facts. Think about mundane facts. Russia is the largest country in the world, spanning eleven time zones over two continents. Russian territory is so large it covers over one-tenth of all the land on Earth, a surface area larger than Pluto. In the Chelyabinsk region of the Ural Mountains, if you stand on the edge of Lake Karachay for just an hour, you'll be exposed to so much radiation from dumped nuclear waste that it will kill you. *Damn.* Why do I keep thinking about Russia? And why with such morbid outcomes?

I step into my dorm room and Lance immediately confronts me. "Aaron came by here earlier in the day looking for you. He's really cray-cray about something. Couldn't make sense out of it, though. I can't believe he's still bent out of shape from you flattening him in front of his girlfriend during that broomball game."

"He deserved that," I defend myself.

"You need to stop bagging him, Joe. What did you do this time that got him so upset?"

I shrug. "Nothing that'll get me kicked out of Carleton." A few papers slide off Lance's desk onto the floor, catching my attention. He has a ridiculous number of printouts and overly large photographs displaying different sections of the night sky strewn all around. "I think you might be going about this the wrong way, Lance, using such a low-tech method on such a high-tech project. It's not a very productive use of your time for the Kepler mission," I comment as I sit down at my desk and start up my computer.

"This isn't for the Planet Hunters' project," he distractedly says as he retrieves the papers from the floor and puts them back on his desk. Instead of immersing himself in whatever this new "project" is, he starts distractedly pacing our small room. After four circuits, he halts in front of me. "What exactly did you do to Aaron?"

His voice is sterner this time, so I give him the detailed answer he's decided he needs to know. "I facilitated killing him in that ridiculous assassin game he plays. It was quite satisfying seeing the expression on his face when he realized I was the one who foiled his elaborate scheme of victory."

"You joined the Assassins Guild?" Lance asks incredulously.

"No!" I shake my head in amazement that he actually thinks I'd consider doing that.

Lance resumes his night watch patrol for a few additional rounds then interrupts his marching to zone in on me with a disapproving look in his eyes. "How exactly did you kill him?"

"I didn't *kill* him. The Assassins Guild shoots each other with Nerf guns, stabs each other with plastic sabers and poisons each other with Tabasco sauce, not exactly life-threatening actions."

"Why did you risk pissing him off again?" Lance recommences his pacing. He's going to wear a hole in the floor before this trimester is finished.

"It seemed like a good idea at the time," I answer. And it was.

"You better hope Aaron doesn't know *you* have a girlfriend now."

"I don't have a girlfriend." At least not here. I focus my attention on the computer and attempt to finish my English essay. Not happening, not with thoughts of Lina and Anna thrashing around inside my brain.

Don't screw up your life again, Joe.

Chapter 11

Tatiana, age twelve

I need my mama's help to figure out why Father's always getting so angry with me, but I've searched through all of her things that he tucked away in the basement and haven't found anything useful. No diary, no journal, no poems or paintings or anything else that could aid me. I'll have to seek guidance somewhere else.

My twelve-year-old brain tackles the challenge and determines that professional help is required. The answer—A Roma Gypsy, more specifically a fortune teller. I know there's one in the local market because I heard some girls at school talking about her.

It's not hard to slip away from the house since my brothers don't keep that close of an eye on me anymore. I bring what's left of my allowance with me, along with one of my mama's favorite silk scarves, the pink one with red roses stitched all over. It doesn't take long to spot her sitting outside a small tent-like structure, wearing a very colorful dress and a deep burgundy bandana.

Striding over, I tell her my dilemma and she beckons me inside. In exchange for my money and the scarf, she agrees to "contact" my mama. We sit down facing each other in two old, wooden chairs. Clutching my mother's scarf in her lap, she closes her eyes. Her body tenses, then she exclaims that she sees the image of a rose. No surprise there, she's holding a scarf covered with them. Suddenly, a grimace fills her face and she shudders. "The flower is wilted and dead."

I feel like I'm punched in the gut. My mama's gone, totally gone.

"I feel your deep pain, child." She opens her eyes and hands me back the scarf. "Bring me something else of hers, and I will try again to touch her spirit. The same amount of payment will be sufficient."

I don't have any more money or anything she might think is valuable ... except the Silver Bear diamonds that Grandpapa secretly gave me. He said to only use them when I really, really need to. Well, this is an emergency of sorts because papa's so mad with me he threatened all kinds of prison-like regulations if I don't act appropriately. Therefore, I think this is a time when I can use one.

Next week, I return with one of the Silver Bears and my mama's favorite decorative hair clip. During this session, the fortune teller gently attaches the hair piece to my braid, then grasps both my hands in hers. "I see a large window with blackness where there should be stars." Seconds later, her body contorts with pain and she's crushing my fingers. "The rose," she shouts. "The thorns are piercing you and the blood is dripping off your hands." Her voice is twisted, very unlike her earlier tone.

I quickly wrench my hands away. What a wretched remembrance! Or is it a cursed premonition? Grandpapa's right. Anything associated with that blasted diamond Mir mine brings nothing but trouble. I shouldn't have used one of the Silver Bears.

Tatiana
Monday, October 1

My first study session with Joe was highly productive and I'm anticipating today's chemistry exam with muted excitement. As

I start working on it, my concentration is disrupted by the turmoil whirling within Joe. I toil my way through the test, taking the entire period to complete it, yet amazingly, I think I did well.

After turning in the exam, I exit the classroom with Ms. Mizuikova trailing centimeters behind me. I'm getting so sick of her prying eyes and pompous attitude.

"Greetingssss Medussssa," Joe hisses at her as we exit the classroom. I'm amazed he's still here because he finished at least twenty minutes before me. "The Greeksss believed the blood which fell from the gorgon Medusssa's beheading ssspawned the ssserpents that now infessst the world. What will happen if your blood is ssspilled?"

She tenses up, like a teakettle getting ready to shrill, and her boiling fury builds to a level where it starts seething out of her pores. Her reaction puts a delighted smirk on Joe's face, which further incites her.

If looks could kill, I'd be witnessing a murder right now. Stanzas from "The Good, the Bad, and the Ugly" start strumming through my head, and I step in to intervene. "I vill see yu at history class, Ms. Mizuikova."

She looms there, undeterred. A student walking by, deep in the throes of texting and not paying any attention to where he's going, jostles into her. The jolt makes Ms. Mizuikova recognize that this is too public a place for causing a commotion. With great difficulty, she backs down and stomps off. I know she won't go far, but at least she's out of earshot.

"Stop antagonizing her," I warn Joe, switching our conversation to Russian. It's much easier to communicate in my native tongue, and there's the added benefit of privacy; not many people around here speak Russian.

He effortlessly follows my lead and changes languages. "I was only informing her of the evolutionary process."

"Well I'm informing you otherwise. Stop it, your people skills suck."

"Coming from you, Anna, that's high praise."

The tone of his voice is indiscriminate. Is he teasing me or trying to upset me? I can't tell. "Stop provoking her!" I repeat. How many times will I have to say this to him? And what is he expecting to accomplish by getting her so wound up? Tapping my neck a few times with my flat-open hand, I plaster him with a huge frown. In Russia, the gesture means, *I am fed up!*

A few students twist their heads and gawk at me, but Joe's eyes shine with a knowing amusement. He's familiar with this gesture, obviously his knowledge of Russia runs deeper than just the language. Why does he know so much about my country? More nagging doubts prod me. I attribute them to my father's constant badgering about never trusting anyone and attempt to push them aside, but too many things about Joe are suspicious, especially his evasiveness and unwillingness to reveal anything about himself.

"How long has the serpent been working for you?" he demands.

Why would he care how long she's worked for me? Stubbornly, we stand there staring each other down while tense moments pass. With each tick of the second hand though, my attraction to him gains more influence, lessening my frustration.

"Aren't you going to ask me how I did on the chemistry exam?" I speak first, redirecting the conversation.

"No."

Ofiyet'! How did we get so turned around? When I saw him, I was eagerly looking forward to discussing the test. Now look at us. What do I say? I don't want to lose the little conciliatory gains of friendship we achieved over the weekend because I don't remember ever feeling such excitement when talking with anybody else. Still, I can't be giving out personal details of my life. However, the date Ms. Mizuikova started translating for me is immaterial, as it doesn't correlate with the date my father hired her. "She's been translating for me since August," I concede.

Joe continues glaring at me. What type of answer was he expecting? Well, two can play at this game. "Why did you get expelled from secondary school?" I parley with one of the numerous questions he has yet to answer.

"None of your business."

Incredibly, amidst this strong undercurrent of animosity, a seductive allure flows between us. Awkwardly, we both try to fend it off, to remain detached and indifferent, but we can't.

"Why did you kiss me, Joe?" I blurt out. Fireworks like an American Fourth of July celebration explode around me, sparking the incredible passion we felt that night outside the pub. Just as rapidly, resentment fires up, then seconds later, mistrust works its way in. *Blin!* My emotions and Joe's emotions combined with Hannah's ridiculous expressions are ricocheting off each other like popcorn in a popcorn machine. How does anyone function this way, with thoughts and feelings all tangled together in a confusing mess?

What did I just ask him? Oh, yeah. "Why did you kiss me?"

Quite unlike my interactions with Viktor, a prior "acquaintance", it's very important to me that Joe kissed me because he wanted to, not because of ulterior motives such as trying to gain my confidence or catch me with my guard down.

"I wish it never happened," he snaps back. "I can never be that person to you, Anna."

What person? A boyfriend? Why can't he be that person to me? I want him to be that person to me, I think. For the first time in my life, I want a boyfriend. But I can't, it would be far too dangerous. "You don't know what I'm thinking, Joe." Secretly, I cross my fingers behind my back. Please don't let this strange bond between us work that way.

"No, but I know what you're feeling, Anna."

Why does that matter? "Feelings are not a rational basis for decision-making," I decisively inform him of this irrefutable fact of life.

He laughs at me. Joe actually laughs at me. I don't understand. A heartbeat later, this double-edged sword of a connection interprets his reaction for me. Feelings *do* determine his actions, or at least some of them, and currently, they're influencing mine as well! The shock of this revelation skews my internal compass as if I were flying over the Kursk Magnetic Anomaly in southwestern Russia. Due to the massive deposits of iron ore there, 30 billion metric tons, magnetic instruments confuse south with east, and north with west.

I no longer have any conception of where true north is! Spinning on my heels, I almost fall flat on my face before Joe catches me. What direction is this relationship going in? Where do I want it to go?

Chapter 12

Tatiana
Tuesday, October 2

The next evening just after seven p.m., our second study session at the library begins. As we start going over the material for tomorrow's class, John Brumba, a student in my kalimba course, saunters up to us, steps right to the edge of the table and intrudes upon our conversation. I can't imagine why. I haven't said one word to him since I've been here at Carleton.

"Anna, I read this wicked article about a Danish band. They loaded a recording studio onto an open boat and launched it across the Arctic Ocean to a Russian city called Pyramiden. It's four hundred freaking miles north of Norway's farthest northern tip, halfway between the Arctic Circle and the North Pole." He squawks this out like an annoying seagull.

Neither Joe nor I deign to say a thing. It doesn't matter. He obliviously screeches on at a whirlwind pace. "Pyramiden's a city without people. The only thing that interfered with the band's recordings was polar bears roaming the streets. Imagine it, using a ghost town for a recording studio. How fleek is that?"

You won't be quite so fleek, whatever that is, if you don't stop pestering me.

Misinterpreting my growing irritation, Joe starts translating for me. "The band used stuff that was left behind there as musical instruments. Their footsteps running on a boardwalk became a drum. Their voices echoing off metal walls became a ghostly choir. Here's the clincher, the really extraordinary twist, are you ready? A fuel tank!" John's eyes bug out as he says this.

Agh. His conversation is even more grating now that I understand all the words. How can I get rid of him without creating a scene? I'm supposed to be keeping a low profile and blending in here at Carleton, not sticking out like a sore thumb. So far, I haven't been succeeding very well with this.

Showing incredible restraint, I clamp my mouth shut and scream at him in my head, willing my inner voice to somehow reach him. *Take a long walk off a short Pyramiden pier.*

"Aren't you going to ask me about it, Anna? It was a huge fuel tank." John theatrically spreads his arms wide, making him appear even more like a pesky gull.

"No," I angrily answer, hoping to discourage him.

"The tank was half full of water and covered in spikes. When tapped, every spike made a different sound. The band used it in their song "Hollow Mountain." Guess what it sounded like?"

Why is Joe still translating? He must be sensing my anger and realizing that I don't want to hear anything else about Pyramiden. Again, I attempt to telepathically direct my inner voice at John. *Go jump into an Arctic lake!*

A group of students enter, claim a couple of tables two spaces down from us and start cluttering it with piles of books, papers and laptops. I watch them try to figure out who is going to sit where. It's rather an interesting process.

A kalimba!" John squawks right in my ear, causing me to jump a few millimeters out of my seat. Astonishingly, he just tweaked my interest. "Vat eez de name of de bandt?" I ask.

"Efterklang," he proudly announces.

"You do know what 'efterklang' means, right?" Joe challenges him. "It's a Danish word for aftersound, signifying that the band takes raw sound, alters it and reassembles it into whatever they want."

Now? Now is when Joe decides to confront him? When I'm finally interested in what he has to say.

"So?" John shrugs him off and starts talking directly at me. "This is so freaking cool, Anna. The total synchronicity of it, taking the kalimba class and then reading this article."

"Vere vaz de—"

"Why don't you go check out the book *Persistent Memories: Pyramiden—A Soviet Mining Town in the High Arctic*?" Joe cuts in. "It's here in this library."

Joe knows about Pyramiden too? That's another subject I've researched that he seems to magically be familiar with.

"Are you messing with me?" John tilts his head, unsure what to think.

"I read it last year," Joe sincerely answers. After a brief hesitation, John takes him at his word and heads off in search of it.

"Why did you send him away just when I wanted to talk with him? And why did you read a book about Pyramiden?" It's certainly not a best seller or standard reading material for college classes.

"We don't have much time, Anna, and we need to study." Picking up a pencil, he begins effortlessly twirling it around his fingers.

How many hours did it take him to learn that? I want to learn how to do that. *T'fu.* It's too easy for Joe to distract me. *Focus, Tatiana.* My gut reaction is to call him out on his avoidance tactic, however, he probably won't tell me why he read about that deserted mining town, and I do need to study.

We resume going over tomorrow's material, but I'm not able to concentrate on chemistry. During my third year of secondary school, I did a project on Pyramiden, and John's references to it are stirring up disturbing snapshots of the town in my head. I tuck my clenched fingers under my chin and fiercely try to pay attention to what Joe's saying, but his words continue to get blurred by my other thoughts. He stops explaining whatever it was he was explaining and looks at me with a questioning expression.

"Pyramiden's such a desolate place," I say. "The isolation, extreme cold and total blackness for months at a time ..." My voice trails off.

"Pyramiden was a Russian's dream, Anna. They were paid double the normal salary and shipped the best food."

"That doesn't mean anything when you have to suffer through endless days of longing and pining for the world you left behind." The amount of introspection in my comment rattles me, and I scrunch my fists tighter. Joe reaches out, gently takes ahold of my hands and settles them on the chemistry book. *Are we switching to osmosis for learning?*

"It wasn't a bad life, Anna" he asserts. "Pyramiden had a concert hall with a grand piano, a library with over fifty thousand books and a heated swimming pool."

"None of that matters. Nothing is good when you're separated a lifetime away," I bitterly state, having no clue where these feeling are coming from. I've never been to Pyramiden and neither has anyone else in my family. "Maps with pictures of airplanes pointing toward home still hang on the walls of the empty houses ... and all the cats died after the town was abandoned. I miss you, Joan."

"Joan?" Joe's face perks up with sudden interest.

"She's every cat I ever had," I lament, missing the purring, the stroking of soft fur and even the occasional tail in my face. I miss them? I don't miss people. Why should I miss cats?

"You gave the same name to every cat?" Joe gives me a baffled look.

"Why Joan?"

"Joan Feynman," I state. "She's an astrophysicist who never gave up. Even when everyone in her life, except for her brother, told her science wasn't a career for women, she still didn't give in. The two of them divided up the universe, and Joan took the auroras."

"I see her appeal." He winks at me.

I silently reminisce about my Joans, wanting to push away the lingering sadness Pyramiden stirred up. One of my cats was a huge orange tabby who loved to stretch out on his back, almost reaching from one end of the couch to the other. He would purr at almost anything. All I had to do was glance in his direction and his Mac truck motor would rev up. The memory brings momentary relief, then my thoughts circle back to Pyramiden. "A tall, metal sunflower marks their common grave," I somberly comment.

"The grave of all the Joans?" Joe's confused.

"No. The grave of all the dead cats in Pyramiden." It's such a travesty. One solitary marker can't come close to recognizing all those loyal companions who were left behind to die.

"Why a sunflower?" he asks.

"I don't know. Maybe because it faces the sun all day, turning itself from east to west." I researched it thoroughly for my school report, but couldn't find a reason.

John reappears from behind a tall shelving unit and flies toward us, firmly clutching a hardcover book. Exuberantly, he holds it high over his head like he's signaling a great victory. As he nears, Joe relays this bit of Pyramiden trivia to him.

"Cats? Huh? Not much of a loss," John casually remarks.

Not much of a loss? All the cats died! Glaring at John, I picture my cat spirit bodily attacking his seagull one, slashing in a valuable lesson. Cats are worthy of respect. *Hmm.* I wonder what type of companion-ego Joe would have. Maybe a polar bear, they live solitary lives in frigid cold and are quite dangerous when provoked. No, that's too pedestrian for him. Maybe a Gobi bear, they exist in extreme conditions, are very persistent creatures and extremely rare. Less than forty-five are known to exist, and none have ever been held in captivity, thus very little is known about them. That's a well-suited match for him.

Strange, my memory lane seems to be stuck in the school projects section. Although, I didn't technically do a report on Gobi Bears because so little is known about them that I couldn't meet the required number of pages.

"Never cared much for cats," John tacks on as he lowers the book in front of me.

No, I imagine you wouldn't. Seagulls tend to abhor cats. "Ya tebe pokazhu gde raki zimuyut," I grumble under my breath.

"What?" John squawks.

"Anna said she wants to show you where lobsters spend the winter. It means—" Quickly, I kick Joe in the shin. He gives me a sly grin, but shuts up.

"Anyway, the photographs in this book of Pyramiden are killin' it," John moves on. "They're so vivid, they draw me right in. It's kinda like I'm inside the picture looking out, seeing the past and present all at once. Creepy, huh? Sorta reminds me of the book *The Golden Compass.* Have ya read it, Anna?" He looks at me expectantly, not realizing I'm having a hard time keeping up with his rapid ramblings now that Joe isn't translating.

"Could the fact that the setting for the book is on the east coast of Spitsbergen on the archipelago Svalbard, the *same* location as Pyramiden, have anything to do with your 'killin' it' connection?" Joe's sarcasm is hard to miss, even for me.

"You've read the book?" John cautiously asks.

Of course he has. If it's anything's remotely connected to me, Joe seems to know about it.

" 'He took the golden compasses, prepared in God's eternal store, to circumscribe this universe, and all created things.' Book seven, page 231, lines 224-226," Joe readily recites.

"There was no book seven. It was a trilogy," John contradicts him.

"I was quoting John Milton's *Paradise Lost.* That's where the American title, *The Golden Compass,* came from. In Europe, it's called *Northern Lights.*"

"Northern lights? Like the aurora borealis?" I excitedly interrupt in Russian. I want to hear about this. "Tell me about the book, Joe."

"It involves a girl, a boy, daemons, parallel worlds and dark matter. The girl seeks to understand the scientific mysteries she encounters, but she's unaware of the critical role she holds in the future fate of their worlds," he succinctly summarizes.

An eerie shiver runs down my spine. More coincidences. "What's the boy's role in this book, and what creatures do the boy and the girl have for their daemons?" I anxiously ask.

John's becoming impatient, not understanding our Russian conversation. "Anna, I wanna show you some of the photos. They're so incredibly wild. You're gonna feel like you're *right* there in this abandoned town, breathing in the salt air and feeling the wind whip your face." He adjusts the book squarely in front of me and flips it open to a picture of a deserted school playground: rusted slides, idle swings that almost appear to be rocking and rough ground marked by stark timbers covered in weeping black soil with rusty red streams. It's a frozen moment in time.

The essence of life that once existed there is still palpable to me, and the long-lost suffering spirals its way upward, seeping into me, just like when I was a child in my mama's bedroom staring at the image of the Chernobyl Nuclear Power Plant. The powerful agony overtakes my soul. The misery of hundreds, yet the misery of one.

Bam! Like a shotgun blast, the slamming shut of the book jerks me out of the past and slingshots me back to Carleton library. Joe's vibrant blue eyes, which are intensely peering into mine, are filled with confusion and pain. "John was telling us about the grand piano left behind in the concert hall. It was a Red October," he haltingly says in Russian, trying to help us both regain our footing.

"So, what do ya think, Anna?" John prompts me.

"Mama taught me on Redt Oktober piano. Vasn't goodt piano." Still, it was good enough to spark my lifelong obsession with music.

"Killin' it! Ya had a Red October piano. Was it shaped like the submarine?"

His little quip further irritates Joe, who's struggling even more than I am to rein in his temper. "What day does Russia celebrate *Red October*?" Joe condescendingly demands. "How about the significance of *Red October?* Do you know anything about either of those, John?"

Having no idea why he's being interrogated about Red October, John stands there mutely.

"Before you start throwing around that phrase, you should understand all its connotations. Red October refers to the Bolshevik revolution of 1917, which led to the first communist government in Russia and the first large-scale socialist state in the world. On October 24, Bolshevik Red Guard forces, led by Lenin, began taking over government buildings. The following day, the Winter Palace was captured. So, what day do Russians celebrate *Red October,* John?" Joe's eyes are blazing with a scary fury. He seems to be spoiling for a fight, and he's quite effectively ramping up the animosity between them, pushing it dangerously close to red-lining.

I'd better put a stop to this before he starts throwing punches and attracting the attention of everyone around us. Piecing together enough of Joe's English words, I intercede. "Redt Oktober eez November 7."

Totally ignoring me, Joe renews his attack. "Prior to 1917, Russia used the Julian calendar. It overestimated the length of each year by eleven minutes and fourteen seconds, therefore—"

"I needt du studti chemistry," I cut in. Joe brawling in the library would cause me an inordinate amount of trouble, not the least of which would be spooling Ms. Mizuikova into high warble. As it is, she barely tolerates waiting on the main floor while we study up here. I don't know how she found out about our first tutoring session, but she did, and now she insists on slithering back and forth to the library with me—a behavior I really want to nix. However, I have to pick my battles with my father, and it was hard enough to finagle this little bit of privacy out of him; I'm not going to push my luck.

Whoosh! With one fell swoop, John snatches the book from the table and zooms off. Joe grabs his pen, holds it like a dart and aims it at an invisible bull's-eye on John's back. "Bet I can skewer him from across the room," he says.

"Why did you provoke him like that?" My fists ball up so tightly that my knuckles are turning white and my fingernails are cutting into my palms, causing them to ache. This experiencing emotions crap is highly overrated.

"Because he was so damn wrapped up in himself, he was oblivious to the effect Pyramiden was having on you, Anna." Dropping the pen, Joe reaches out and simultaneously traces intersecting circles over the back of both my hands with his index fingers, causing goose bumps run all the way up my arms.

I pay closer attention to the pattern he's drawing on my skin and notice it's the symbol for infinity. I pull back, alarmed. "Couldn't you have found a different solution?"

"And what would you suggest I have done?"

"You should have ... " *T'fu.* I have no effective strategies on how to handle all these blasted feelings.

Joe leans in and switches to gently massaging my throbbing hands. It's deliciously comforting. "Why did that photo upset you so much?" he asks.

It triggered a bad memory from long ago, but I don't know why. Instead of telling him this, I ask the more disturbing thought that's been circling around me for some tome. "Do you believe things happen for a reason?"

"If you mean do I believe in predetermined fate, no." He states this so emphatically and with such certainty that I begin to vacillate about my own wisp of a notion.

"How can you be so sure?"

"Because there are terrible things in this world, Anna, that should never happen, and they happen anyway."

I attempt to accept his rational assertion, but I can't because on rare occasions there are things that seem too well ordered to be purely coincidental. Take the word "red" for instance, which is currently spinning in my mind due to Joe lecturing on the subject of Red October. Less than two months ago, I typed in the words "Red Planet" and was amazed at what I found.

The next logical step is permanent settlement on Mars, and the Mars One project has been founded with the goal of establishing a permanent settlement on the Red Planet. Excitement builds within me; maybe I can escape this world and all my troubles here. *The mission will place high demands on the first crews. They will be required to fix their own technical and medical problems, grow their own food and harvest water. They will only be able to communicate with friends and family on Earth through current technology with its differing time delays.* I don't see any problems for me there. *Some humans have always been explorers, it's in their genes.* Me, me, me! The stars have been calling to me all my life. *Do you have what it takes to make history, change the course of human evolution and journey to the stars?* Yes, I do. *Mars One is searching for you, the crew of the first Red Planet colonists. Apply online by August 31st at the Mars One website and start the adventure of a lifetime.*

I stumbled on this announcement by chance, just in time to apply before the deadline. It never would have been an option if I was still in Russia. I tilt my head skyward, trying to envision what it would be like to have another chance at life, one where I'm part of something incredible and the dangers that loom around me are totally different than these here on Earth.

It's too soon to get my hopes up though, because fate often has a way of crushing them.

Chapter 13

The Survival of the Russian elite. In the self-serving, corrupt capitalism of Russia that followed the Soviet Union's collapse, a small group of people amassed vast fortunes. These billionaires then used their wealth and power to influence Russian politics. In 2001, following Stalin's example, Vladimir Putin began a brutal campaign to quash these outspoken oligarchs. Those that would not be silent escaped to London or were jailed. The British capital became so flooded with Russian oligarchs that it was dubbed Moscow-on-Thames and Londongrad. In 2006, these exiled Russian oligarchs and their associates began dying under mysterious circumstances.

Vladimir Ivanovich Alliluyev, Tatiana's grandfather, is an exiled Russian oligarch.

Tatiana
Three years ago

"Keep your mouth shut and stop publicly denouncing the president, Tatiana."

"Why, Father? Grandpapa doesn't."

"Your grandfather is a bitter old man who is more concerned about his legacy than his family."

"At least he's doing something, Father."

"Doing what? Disillusioning his friends, putting his family in danger and getting himself targeted for assassination?"

"No, he's—"

"Yes, Tatiana, that's all he's accomplishing."

Tatiana
Two years ago

"Stop screaming at me, Father. I'll stop criticizing the president."

"It's too late for that now. Since you couldn't keep your mouth shut, you've become a problem to them that needs to be eliminated."

"That's ridiculous. They're not going to kill me."

"Open your eyes, Tatiana. Alexander Litvinenko, Arkady Patarkatsishvili, Paul Klebnikov and Sergei Magnitsky. They're all dead because they couldn't keep quiet."

"Those are all men. And anyway, I know how to protect myself. You've made sure of that."

"You're not going to be tolerated because you're a woman. They'll come after you, just like they went after that journalist, Anna Politkovskaya. She was shot twice in the chest and once in the head at point-blank range inside the elevator of her apartment block!" He's yelling louder, like that will make a difference.

"I can take care of myself, Father," I repeat.

"Yes, right into your grave. It's no longer safe for you here. I'm sending you to live somewhere they can't find you."

Tatiana
Wee hours of Wednesday morning, October 3

My throwaway phone is vibrating. Nothing good can come from

this. I try to ignore it, but the continued buzzing aggravates me into answering it.

"Hallo," I grumble while attempting to wipe the sleep from my brain.

"Tatiana."

"Yes, Father," I dutifully reply. Couldn't this wait until a sane time of day?

"Ms. Mizuikova forwarded me more information about that Russian who is taking a special interest in you." By the low pitch of his voice, I recognize there's no leeway for skirting around the issue.

"He's not Russian, Father, he's an American student." Or so I think, unless he's working for you.

"I've checked into the matter, and he is not who he claims to be. Ms. Mizuikova reported he speaks fluent Russian, yet all records for a Joseph Earl Carleton show him being from Hibbing, Minnesota, with German parents and no relatives, friends or acquaintances who are even remotely connected to Russia. I had a contact discreetly converse with one of Carleton's brothers, and he reported back that no one in the Carleton family knows how to speak Russian. Stay away from him until I can look into this further."

My father seems to be taking this very seriously, perhaps he didn't hire Joe. "I'll be on my guard at all times." I feed him one of the rote answers I use to placate him. It doesn't fool him this time.

"Listen to me, Tatiana. Joseph Carleton was a below-average student in secondary school who ranked in the bottom twenty percent of his class and was expelled during his senior year for stealing final exams in a cheating scam. He would not have been accepted at Carleton College. There's something more going on here. Stay away from him." My father's tone is stern and uncompromising, too uncompromising for it to be a ruse. He didn't hire Joe to watch over me.

I don't respond and he repeats his decree. However this time, there's a touch of pleading in his voice. "Tatiana, keep away from him."

I don't have the capacity to outright lie to my father, and having had only a few hours of sleep, I don't have the energy to argue, but I refuse to let him dictate my every action from a continent away. Shutting off the burner phone, I stick it in the trash can next to my bed. Later in the morning, I'll get another one out of the locked drawer in my desk. My father made sure I have a ready supply on hand.

For the remaining early-morning hours, I toss and turn in a restless, semiconscious, in-and-out of sleep state. At least I'm not having that dream about Mars and the old woman prophesying to me that *I'm the one*. Ever since Joe kissed me that night outside the pub, the dream has disappeared. Is Joe the one I'm supposed to help?

I puzzle it around for a bit, then decide to get up out of bed. My father's damning accusations have made me anxious to question Joe further. As I dress, I'm methodically thinking of how to go about this. Obviously, it needs to be done without Ms. Mizuikova around. I quietly unlock the door and slip into the hallway. That's when I finally realize I don't know which dorm Joe lives in. I don't know his cell phone number and I have no idea how to contact him without bringing undo attention to myself. My father's recent words of admonishment start wriggling more worms of doubt into me. *Tatiana, he's not who he claims to be. Stay away from him.*

No! No! No! I won't let my father do this to me again. I'm more than capable of making my own decisions. Stepping back into my room, I begin reformulating my plan. *Just great.* Hannah wakes up, all bubbly and cheerful, and begins rummaging around the room to gather her things. Before leaving for her daily ritual, which requires bringing most of her personal toiletries and accouterments to the bathroom with her, she optimistically asks, "Anna, do you want to go shopping with me this afternoon?"

"No." This recently resumed, continuous pestering for us to be buddies is quite a nuisance.

Hannah reluctantly trudges out of the room and plods down the corridor. When I no longer hear her footsteps, I close the door and start reading the book about the Russian space program my father gave me on my last birthday.

A mere eleven short minutes later, the door flies open and Hannah barges in, noisily clomping around, seemingly searching for something. "Anna, I forgot my waffle maker and my electric chain saw."

What now? What outrageous scheme is she employing to get us to become friends? All I wanted was the usual thirty minutes of peace I've grown to expect during her morning routine. Is that too much to ask for? Not bothering to look up, I continue reading.

"They've got to be around here somewhere, Anna. Have you seen them? They're both neon green." Hannah shuffles around more stuff, pulls apart more things and makes an inordinate amount of noise. Without taking my eyes off the page, I vigorously shake my head no. She's making even less sense than normal.

"Anna, my hair's on fire!" She screeches at near-maximum volume, causing me to drop the book, clamp my hands over my ears and take cover. When her tortuous attack on my eardrums ends, I concede. *Okay, Hannah, I give up.* Letting out a frustrated sigh, I glance over at her and my jaw instantly drops open in shock. Her hair is flaming pink and spiked out in every direction, her eyes are literally glowing yellow and her outfit is a dizzyingly iridescent, spiral pattern.

"It's about time you looked up," she says. "I was running out of ideas." Using theatrical dance movements, she waltzes over and fans her fingers out in front of my face. Each one is decked out with a huge ring containing a stone of a different color.

"Vat are you doing?" While my behavior may be bizarre at times, hers is not even in the same galaxy.

"I'm an anime character, Anna. It's my Halloween costume from last year's party." She spins around like a circus clown, her shower shoes sticking to the floor.

"But vy are yu duing diz?"

"It's a sad state of affairs when *this* is what it takes to get you to talk with me. So, how about we go shopping?" she urges.

"No." Grabbing my backpack, I leave to find somewhere else to read. The library should be a quiet place. Upon reaching the entrance, I pull out the cell phone I use for local communications and begrudgingly inform Ms. Mizuikova of my change in whereabouts. No need to stir her up needlessly. Not being where I *should* be would likely cause more of my father's stringent security procedures to be implemented.

I manage to read one chapter before needing to depart for chemistry. During class, I bide my time by sorting through conflicting assessments of Joe. Yes, there's mistrust, however, there's no malice. In fact, there's even a faint, mutual solicitude between us. While Joe may be misleading me, and there's certainly a great deal he's not telling me, I don't believe he intends to harm me. Besides, there's something my father's overlooking. This is Joe's second year attending Carleton, a fact confirmed by Hannah and numerous sports articles written about his football feats. Therefore, since he started attending school here before I even sent in my application, there's no way he could have known back then that I would be at Carleton now.

At long last class ends. Students begin conversing with each other, pulling out cell phones or approaching the professor to ask questions. Joe remains seated with a complete lack of interaction. We're not so different, he and I. We're both displaced puzzle pieces that long ago were relegated to the wrong box and no longer fit into the larger framework.

"Anna, why are you still sitting there?" Ms. Mizuikova gives me her usual scoff.

To count frog hairs! I suddenly see the benefit of Hannah's fifty-foot pole—keeping certain people far, far away. Wish I had one of those beauties in my hands right now so I could order Ms. Mizuikova out of my life, or at least fifty feet out, and force her to comply.

Accidentally on purpose, I push her aside as I venture toward Joe. He doesn't move an iota. It's not until I reach his side and provoking words start pouring out of his mouth that I realize my shortsightedness; he was baiting me. These bothersome, conflicting emotions are so distracting.

"I don't dare look up toward the creature with the hideous face and venomous vipers in place of hair. Legend says that just to behold the sight of Medusa turns a mere mortal man to stone," he snipes in Russian while mockingly shading his eyes.

Why must he continue harassing Ms. Mizuikova? The result is a sideways domino effect, making it increasingly harder for me to manage her. How can I get this across to him?

"I need to talk with Joe privately," I firmly direct Ms. Mizuikova, hoping my father had some sort of brain fart and didn't relay to her the information about Joe he told me earlier in the morning. The beyond-menacing look on her face confirms that she is armed with this knowledge, primed and cocked, and ready to go for his jugular.

"Wait outside the classroom." I sternly order.

"I will not do that!" she angrily snaps back. "Your father instructed me to ensure you do not go near him until he's able to investigate—" Her half-finished authoritarian statement hangs in the air, implying details about her position and my father that should not be known; she's much more than just my translator, and my father has the capability to delve deeply into matters.

Glancing at Joe, I see there's no chance he overlooked this. Ms.

Mizuikova's careless actions just increased the animosity he already harbors toward her *and* added in an element of suspicion.

"It would not be in your best interests," she awkwardly restates, trying to cover her tracks.

"Go, now." I point toward the exit.

With a loud stomp, she sluggishly moves just outside the classroom, keeping the door wide open so there's a clear view of us. No doubt she'll be informing my father of this situation pronto. I have to hurry.

"Do not get her more riled up," I confront Joe. Although I don't know if that's even possible at this point. "There will be serious repercussions for both of us." I'm sure my father's intensely scrutinizing Joe's identity as we speak.

He sets his shoulders back, rises out of the chair, stands tall and rigid as if he's at attention and gives me a mock salute. "Aye, aye, Captain Anna. Damn the torpedoes, four bells ahead." His words are clipped, short, and loud, more than loud enough to reach Ms. Mizuikova.

No, no, no. Doesn't he have a disengage program or default mechanism wired into him? Why is confrontational his only communication mode with her? And why is he so upset anyway? He's the one who instigated this whole situation.

"Remember when I told you it would be a bad thing if my father knows more about you than I do? Well that has happened, and we need to deal with it."

Snapping his heels together, he gives me another salute. "Understood, ma'am, but *I* am not in your chain of command."

I don't have time for this. "Tell me why you were expelled from secondary school? I have to know so I can get my father off my back." If I can't confirm who Joe really is and why he's here at Carleton, I have no chance of warding off my father. And if he finds out anything he deems "life-threatening," I'll be forced to move again. I don't want that.

"Does the serpent always slither after you every moment of the day?" He slithers his hands through the air in a most annoying manner.

I'm not letting you slink us on to a different topic, Joe. "Why were you expelled?"

"So, since my father doesn't take an irrational interest in my acquaintances, you don't have to answer my questions?" he inquires.

"Why is this situation so difficult for you to understand?" I explode. That didn't take long. Joe's quite adept at activating my trigger points. I take a few deep breaths to regain some control. Alright, direct demands aren't working, a different approach is necessary. "Pozhalvista," I politely appeal.

Saying please causes a gentle exchange to pass between us, and amazingly, Joe extends a truce. "We know almost nothing about each other, Anna. Can we at least attempt a normal conversation before you march me down the long, dark alley of fact-finding?"

I'm caught off guard by his request and scrutinize it for ulterior motives. He appears sincere, yet I'm leery of his intentions since none of my doubts about his identity have been resolved. "What type of normal conversation?" I hesitantly ask.

"Let's find a better spot to talk."

When I nod my agreement, Joe heads out into the hallway and I follow. That's weird, there's no sign of Ms. Mizuikova. Why would she disappear like that and what is she up to? Exiting the building, we walk off a ways then duck behind Watson Hall and enter a Japanese garden that I didn't know existed. Following a path of sunken granite slabs, we arrive at a curved, redwood bench.

"What do you miss?" Joe promptly asks as we sit down, his Siberian blue eyes glittering with a strange intensity.

My mind glides over his question and "My mama," slips from my lips before I can stop it. I've never voluntarily allowed myself to dip into this well of loss. Barely surviving the aftermath of her death, I vowed to never go near those lethal feelings again. *Nooooo!* A stifling, dark cloud of horrifying anguish begins draping over me, pulling me down into that terrifying pit of buried agony. I can't stop it ... I can't see ... I can't breathe!

Joe instantly pulls me close. His strength radiates into me, staunching the rising tide of panic. My gasping breaths ease and my vision begins to clear, but the traumatic ordeal weighs heavy over both of us. Contrary to what I would have expected, this heaviness brings us ever so slightly together and loosens ever so lightly the wall of ice I've been hiding behind.

"What was she like?" Joe asks.

"Willful, tall, formidable," I whisper. *Tatiana, you can't talk about your family.* "My mama protected and sheltered me, she was my rock to hold on to in a storm," I continue. Responding to him seems out of my control as the memories I've kept buried for a lifetime start rifling through me. I smile as fond recollections surface. It's so wonderful to quickly dart among them. Unfortunately, the flip side will not be denied, and the petrifying fears begin to arise as well. Firmly clenching Joe, I try to will him into conquering my demons.

"Tell me about your father," he softly prompts.

Is this the way an ordinary conversation is supposed to go? Does he have an ulterior reason for asking these details about my parents? Once again, my father's early-morning warning buzzes through my head. *Tatiana, he's not who he claims to be.* Could Joe be a sleeper agent? *Don't be ridiculous, Tatiana, the Russian regime wouldn't waste such a valuable resource on you; besides, they already got what they wanted. You're silenced and out of public view.*

"Why is your father so overprotective of you, Anna?"

A myriad of competing emotions pull at me. It's a tangled,

dangerous web Joe and I are weaving. Dangling between strands of skepticism and acceptance, wanting to learn more about each other yet unwilling to reveal much about ourselves, we cautiously advance, draw back, then repeat the cycle again. What should I do? Which of the conflicting signals do I follow?

My emotions make the decision for me. Fearing the terrifying abyss of my tormented memories more than my father's wrath, I do the unthinkable. I start telling Joe about him. "When my mama died, my father sank into a deep depression. It was only by completely submerging himself in his work that he managed to cope with the loss. From then on, he ignored me for most of my life, until recently. Now he's continuously worried and fixated on my safety."

"What about the rest of your family, brothers, sisters, grandparents, aunts, uncles, didn't anyone help you deal with the death of your mother?" Joe appears confused by this.

I'm perplexed. Why didn't he ask me more about my father or why he thinks I'm in danger? "My brothers are a good deal older than me and had problems of their own to deal with," I explain. "Also, when Father was home, they spent the majority of their time talking with him and discussing things he cared about like politics and history. Neither of them paid much attention to me."

"What's your father's work?" Joe makes this sound like an innocuous question, but his face betrays a keen interest. More alarm bells go off in my head, a little louder this time. Maybe Joe is a sleeper agent, it's just he was placed here for a totally different reason and stumbled upon me by blind luck. *Ofiyet'!* I'm becoming as paranoid as my father.

"Does he work for the Kremlin?" Joe prods.

"No!" I snap into full awareness. What have I done? My father will be furious if he finds out I've told Joe anything about our family. I can already hear him screaming at me, *"No details of your former life can be revealed!"* I start biting my lower lip in frustration, staring out over the bowed, wooden bridge that spans the Japanese garden's "lake" of rocks.

" 'Every man has his secret sorrows, which the world knows not; and often times we call a man cold when he is only sad,' " Joe recites in a somber tone. "Henry Wadsworth Longfellow sure nailed our predicament, didn't he, Anna?"

Releasing the death grip on Joe that I didn't realize I still had, I scrutinize him closely. The pain he constantly suffers from is becoming clearer to me, because it's a pain I'm suffering from too, an unremitting sense of sad aloneness. Our isolation is further exacerbated by the inability of the world around us to comprehend it.

"Yes, Joe, he did." Suddenly, I'm extremely curious to hear his reply to a "normal" conversational question. "What do you miss?"

"I miss Penny," he instantaneously responds.

I wait for more particulars, for a tiny glimpse into his inner self, but nothing more is forthcoming. "That's it? Who's Penny?"

His face brightens with unfettered delight which shines out of his eyes and shimmers around his smile. "She's chock full of fire and brimstone and always happy. She's faithful, loyal, obedient, protective and my exuberant supporter."

"It sounds like you're describing a dog," I remark.

"Anna, you are to stop talking with him immediately!" Ms. Mizuikova yells across the expanse as she charges into our secluded oasis.

How did she find me here? Either she's been tracking my cell phone or she's been spying on us this whole time. Either way, my father will hear about this. I've had it with her overbearing intrusions in my private life.

"I do not take orders from you, Ms. Mizuikova, and I'm going to get you fired. You're a ticking time bomb that is out of control." I can't fathom why my father hired her.

Ms. Mizuikova's index finger threateningly points at Joe, then me, then Joe, then me, then Joe, until the realization sinks into her head that my threat may have some merit. Emitting a deep growl, she retreats to a nearby pagoda statue, pulls out her phone and starts taking pictures of us. My window of opportunity is becoming extremely limited.

I refocus on Joe. There's a hint of admiration in his expression. He'd be thrilled, maybe even more than me, if she got sacked. Hopefully, that will happen very soon. "It's still your turn, Joe. I disclosed personal details about my life. Talking about your dog is not reciprocating. What do you miss?" *Please give me some insight into your life.*

"She is what I miss, Anna. She's a Catahoula Cur with a cocoa fur coat that's dotted with deep-mocha spots, except for the tips of her paws, which are white. Whenever she sees me, her golden eyes light up with pure joy. Penny's soul isn't bound by human constraints." His mood is becoming extremely upbeat. Joe seems to exist wholly within his feelings, his world dominated by them, while I exist totally outside my feelings, my world devoid of them ... or it was until I met . And since then, I've had three cataclysmic meltdowns.

Contemplating his answer, I feel cheated of the opportunity to learn more intimate information about him. However, this is the one time I've seen him anywhere close to happy, and I can't deny him that. "Tell me more about Penny."

"She can leap four feet straight up off the ground from a standstill. During winter break, when I'm home sitting at the kitchen table, I can see her head bobbing up and down outside the bay window as she pantomimes to me, 'Hey, remember me? The one out here in thirty degrees below zero.' " An utterly enchanting grin emerges and I smile back, enjoying his sharing of memories. It feels good.

Heartened by my reaction, Joe continues with greater enthusiasm. "Penny constantly prances about with a massive stick in her mouth. Sometimes it's so large she can barely lift it. When left alone, she chews and shreds branches all over the front steps, making my sister Beth complain that a beaver's taken up residence. Penny lives like she owns the world. That's why I nicknamed her Princess."

His boyish passion is so contagious that droplets of it effortlessly pass through my skin and spread throughout my body, making me giddy with happiness. Fervently, we soak up this rare bliss just like residents of Russia's largest city north of the Arctic circle, Murmansk, when they soak up light after forty days of no sun.

Is this what it's like? Is this what it's like when your soul sings?

"A friend opens a place hidden deep inside of us, and only then can a new reality happen," Joe remarks.

Is he still referring to his dog? He changes subjects so abruptly, I often can't keep up. "Is that a Kepler quote?"

"No, Anna, it's us." The strange energy rippling between us sends out an added burst.

"I just finished talking with your father," Ms. Mizuikova shouts as she rushes down the path toward us. "You are to come with me immediately."

No more time left for chatting. "Why did you get expelled from secondary school, Joe? Please tell me."

He shifts his eyes toward Ms. Mizuikova, then back at me. "I was caught with some copies of final exams."

Incredible, a straightforward answer. I know it's true, because it corresponds with what my father told me *and* Joe's not feeling guilty or anxious. "Were you going to use them or sell them?"

"I wouldn't do that!" he snaps.

"Anna! You—*oomph*!" In her haste, Ms. Mizuikova trips over one of the granite stones and face-plants, sprawling upon the path and giving me a few extra moments.

"Why did you steal them? I have to know, otherwise my father will never allow me near you again." *And I don't want to be taken away from you. I like you.*

The edges of Joe's face harden as my ultimatum tests our tenuous camaraderie. Then his stern expression softens and he confesses, "I wasn't the one who took them."

"Then who did and why did you have them?" I'm desperate and I'm running out of time.

"Now!" Ms. Mizuikova grabs my arm and starts trying to pull me away. She actually yanks hard enough to bring me to my feet.

"Is there a problem here?" a gravelly voice breaks in, speaking English.

We all turn our heads in unison to see an elderly man standing near some bushes, holding gardening tools and staring at us. Where did he come from? I've got to pay better attention, but that seems to be impossible when I'm around Joe.

Analyze the situation, Tatiana. The man's back is dipped with age, his gray hair is barely visible under his khaki sunhat, his skin is leathery and wrinkled, his thick eyebrows are deeply furrowed and his ankles have braces on them. I determine he's exactly what he appears to be, a gardener, and not a direct danger to me.

"No/Yez," Ms. Mizuikova and I reply simultaneously. He's not sure who to believe. Having a witness who seems willing to intervene gives me the advantage. I boldly plant my feet and threaten Ms. Mizuikova. "Let go of me and leave us alone, or I'll get campus security involved."

She fumes back toward the entrance of the garden and again pulls out her cell phone. The elderly man winks at me then resumes his task of trimming the shrubbery. Problem solved for the moment. I quickly return to the issue at hand. "Why did you have the exams?"

"I've already been questioned about this many times, Anna. I didn't answer before and I'm not answering now."

The ludicrous action of turning him over to Ms. Mizuikova for interrogation crosses my mind for a brief instant. I brush aside that ridiculous idea and a burst of insight hits me. "You're protecting someone," I state with certainty.

"Zhit' ne polzhi," Joe grumbles under his breath. He stares at the ground and starts kicking some pebbles around.

Live not by lies? My grandpapa mumbles that all the time. You might even say it's his mantra. Quite an unlikely coincidence. Maybe my grandpapa hired Joe. "Who are you protecting?" I demand.

He brusquely gets up and starts walking away. Racing after, I follow him out of the garden and onto the sidewalk. Laboriously I wait, determined to be the victor in this contest of wills. But the longer it continues, the more Joe's inner conflict cuts into me and chips away at my resilience. *Guilt,* isolation, *guilt,* regret, *guilt,* heartbreak, *guilt, guilt, guilt.*

I hold up the white flag. I don't care anymore whether he answers my questions, and I don't care anymore about justifying his actions to my father. I won't be the source of his increased suffering. Hopping in front of Joe and forcing him to stop, I reach out and hesitantly run my fingers along his scars, down his cheekbone, over his shoulder and finally rest them on his heart. The gentle gesture creates a feeling of tolerance and acceptance between us, two things that have been sorely lacking in our lives. "You don't need to tell me, Joe."

He takes ahold of my hand and interlaces our fingers. "His parents got in a car accident, and his mom was killed. Afterward, his dad started drinking, the bills piled up and then his sister ran away. All he wanted was to get some money to help his family. I couldn't let him take the fall for that." Joe's swirling blue eyes are becoming moist around the edges and a sad grimace flits across his face as he wrestles with regrets.

"So you took the fall instead?" I ask.

"Everyone already expected something like that from me."

I try to understand, but something's not adding up; it doesn't *feel* right. *Wow!* These blasted emotions can actually be useful. "What was the name of your friend?" I inquire, then immediately feel ashamed of my phishing attempt.

"Do you wanna get some food?" Joe deftly deflects my question.

I want more answers, but he's not going to give me any more, at least not right now. I accept his offer. We walk hand in hand and Joe's mood improves, even with Ms. Mizuikova trailing behind us.

"The viper's keeping an especially close eye on you today," he comments.

"I already told you, my father looked into your background."

"And what did he find?"

"You never explained why you didn't want to speak Russian that night at the Contented Cow Pub." *I can evade questions, too, Joe.* "Why were you so worried about speaking Russian? What difference does it make if anyone knows?"

"I flunked an impromptu Russian quiz and didn't want the professor finding out that I speak fluent Russian." He gives a half shrug.

"How could you possibly have failed a Russian quiz?" Obviously, he flunked it on purpose.

"My high school coach believes I have dyslexia." I shoot him a scowling I-don't-buy-it look, and it persuades him to give me a different answer. "It's complicated."

"That's not a complicated question. Here, I'll give you an easier one. When did you learn to speak Russian?" He must have learned at a young age in order to speak so fluently with only an occasional hint of an accent.

"A while ago."

What a vague, safe and totally useless answer. "From whom did you learn to speak Russian?"

"I learned by listening and practicing," he skillfully sidesteps.

"Why is it that no one else in your family knows Russian?" I press him, picking up the cadence of our back-and-forth repartee.

He eyes me watchfully, then replies in a measured tone, "No one else wanted to learn."

I fire off my next question even faster, keeping him on the defensive and hoping to prevent him from turning the conversation elsewhere. "Why does no one else in your family know that *you* speak Russian?"

This time, he's caught off guard and hesitates. "Who says they don't?"

"Who taught you Russian, Joe? Why is that such a secret?" This is so frustrating. Viktor would tell me anything I wanted to know, even confidential information about his military service. His loyalty was to me, not the corrupt government.

"Your father?" He looks at me askance.

"My father did not teach you Russian." Now he's just answering with total nonsense to dissuade me from asking more questions.

"No. I'm asking if your father is the one directing this interrogation."

"Just tell me what you're hiding. It'll be better for both of us that way because no matter how well you think you've concealed your secrets, my father will eventually uncover them." Or at least ninety-five percent of them. I learned that unfortunate fact about my parental unit at an early age. Viktor was one of the rare things in my life he never discovered.

"No, he won't, Anna," Joe tersely states. The clenching of his jaw muscles and the slight tightening of his grip around my fingers relays the rest of his message loud and clear. Back off.

A group of female students pass by and gape at us holding hands. *What's that all about?* We reach the east dining hall and Joe opens the door for me, keeping it ajar just long enough for us to whisk through. When Ms. Mizuikova steps forward, he soundly shuts it in her face. Giving me a sly wink, he leads us to the serving area. Joe selects a grilled chicken sandwich, french fries, a side salad and chocolate milk. I decide on the same, I'm not in the mood to be wasting time with mundane decisions.

As we wander over to a table, I notice countless sets of eyes on me. In fact, most of the students in the dining hall are watching with a disproportionate level of interest. When we sit down together, they all gawk with disbelief. What in the world did I do to be in the spotlight this time? Or is a residual effect from Saturday night's debacle in the pub? Or maybe it's not me they're fixated on, it could be Joe.

"Does this happen to you every day?" I ask as I take the food off my tray and organize it in front of me.

"Does what happen every day?" He starts devouring his lunch, seemingly oblivious to the rapt attention of everyone in here.

"Do all these other students always blatantly stare at you?"

"Not usually," he garbles between bites.

"Then why are they doing it today?" I'm tempted to get up and leave, but it's too late for that. My every action is being monitored by a battalion of strangers.

"Because you're sitting with me." There's a touch of amusement in his voice, and he actually stops eating for a moment to smirk at me.

"And why am I accorded this special honor?"

"Because you, Anna, are the first girl they've seen me pay any attention to since I came to Carleton."

Chapter 14

Joe, age nine

I force myself to walk up the steps and into Kluge's house. I can't believe nothing has worked yet. I can't believe he hasn't kicked me out yet. What's it gonna take?

Leaving my backpack by the front door, I sneak farther into the house. Peeking into Kluge's back room, I spot him sitting at his writing desk. He's mumbling under his breath in words I can't understand. Suddenly, I know exactly what to do. Creeping up behind him, I scream in his ear as loud as I can. "Yeban'ko maloletnee, sookin syn, Svoloch." For once in my life, I'm glad I have a perfect memory.

Having his own angry words spat back at him does the trick. He flies out of the chair, grabs me by the collar, and raises his hand to strike me ... but he doesn't. Instead, he just looks down at me as tears form in the corners of his eyes. "Niemand hat seit über vierzig Jahren auf Russisch mit mir gesprochen," he confides in German.

No one has spoken to him in Russian for forty years? What the heck? Who is this old geezer?

The sad look lasts only a few seconds, then his mean-old-man look returns. Twisting my collar tighter, he bends over and puts his face right up next to mine. "Man kann nie jemand darüber erzählen, nie.

Verstehst du mich?" he orders.

I nod yes. I won't ever tell anyone you speak Russian.

Joe, age twelve

Kluge doesn't know I learned how to read Russian, but he won't let me look at the letter he's been writing anyway. What can possibly take so long to say? He's been working on it for over two weeks.

As quietly as I can, I set the water and medicine on the side of the desk. Not quietly enough. Kluge hears me and quickly turns over the piece of paper, but not before I see a few more words.

I have kept the silver bears hidden all these years.

Silver bears? What are silver bears?

That evening, when I have a chance to get on a computer, I start searching. It takes a while, but I finally find an answer. The Mir diamond mine in Siberia. It's the largest open-pit diamond mine in the world, and during its peak, the mine produced over ten million carats of diamonds annually. Two million were gem-quality diamonds dubbed Silver Bears.

Joe
Thursday afternoon, October 4

Anna's father will never discover my secrets. I've been telling myself this since yesterday afternoon.

No one else in the world knows what I know now that Kluge is dead, and I'm sure as hell not going to inform her father or anyone else about it. But how has he found out so much about me? My "theft" of those final exams isn't public knowledge, yet Anna wasn't the least bit surprised when I told her about it, because she already knew. And how did he find out that no one else in my family speaks Russian?

Kluge's stern warning resounds in my head, *know your enemies, Joseph*.

I need to do some digging of my own to figure out just who her father is and how he has the resources to dig so deeply into my life. But where do I start? I have almost no information to go on, for I'm fairly certain Anna Alkaeva is not Anna's real name.

My attention shifts to this bizarre link between us that's gone way beyond normal. At first, I rationalized it away. Most people don't realize how many nonverbal cues they give off—body movements, postures, gestures, rates of speech, inflections, facial expressions. With my eidetic memory, I can recall all of that for every person I've met, often allowing me to accurately predict their reactions. Not so with Anna, she's very unpredictable. Also, I assumed mirror neurons were playing a role in our emotional connection. However, all of these things are contingent upon me actually looking at her, and I'm experiencing her feelings even when I'm not looking at her. I can't make any sense of it.

Walking into chemistry lab, I'm still consumed with this weird "entanglement" Anna and I are stuck in. I'm busy trying to figure a way to skew the variables enough to break free of it when Aaron smashes into me.

"Sorry, didn't see you there," he says in a sickly polite voice. His perfectly groomed hair and wrinkle-free, button-down, preppy shirt give him such a "sophisticated" look. With difficulty, I refrain from beating the crap out of him.

"Someone else who appreciates your charms, I see," the Gorgon Medusa hisses at me in Russian as I approach the lab table. She's wearing one of the three pantsuit outfits she rotates through. There's black, steel grey and dark navy blue.

"I thought you got axed yesterday. Why are you still slithering around here?" I heckle her in English. My other new lab partner, Trisha, covers her mouth to smother a laugh.

"Stop it," Anna cautions me in Russian. "Go get the flasks we need."

"Not before donning my gorgoneion shield. Gorgons forbid I don't protect myself from the monstrous dangers that—" Aaron slams into me a second time, then shoves his way past toward the flasks.

"Not so tough now, are you, Carleton?" he delightedly taunts over his shoulder.

"Muhahaha." Medusa delightedly jeers at me with her odious laugh, making my gut churn. I'm gaining a lot more admiration for Anna, being able to put up with this woman.

"If I were you, I'd deck that guy," Anna counsels me, sharing advice from her extremely limited repertoire of people skills.

I wonder if she actually went around physically pommeling people in Russia? Maybe that's why her father sent her to the US, as a kind of reform school to curtail that behavior. The stakes are far higher for her here than back in her homeland, where street fights would probably be overlooked. If she hits someone here, she can get charged with a crime and deported back to her father, somewhere she obviously doesn't want to be. Or maybe she's here to avoid getting sent to prison. Anti-government protests can land you in prison, exiled or eliminated in Russia, and Anna seems to have a hard time keeping her mouth shut.

"No, Anna, if you were me, you wouldn't deck him. 'Cause if I get caught fighting again, I'll get kicked off the football team and probably kicked out of Carleton. Unfortunately, Aaron knows that." I head for the flasks, paying better attention to what's around me. Aaron's purposely lurking nearby, waiting for another chance to harass me. This go-round, I'm better prepared. "Careful, Aaron, improper selection of equipment can cause explosive results."

"You're a real ass, Joe. It makes pummeling you in front of your girlfriend all the more enjoyable." His perfectly white, perfectly even teeth gleam like he's a poster child for an orthodontist's office.

I can fix that condition for you, Aaron. Restricting myself to verbal attacks, I flippantly remark, "Have you memorized the chemical hazard information chart yet? Never know how you might get assassinated next." Adding a little more power to my words, I throw a punch at him, stopping my fist inches from his shining teeth. "Kaboom!"

"Screw you, Joe." He quickly snatches two flasks and bolts away. I grab our flasks and rejoin Anna and Trisha. As we're nearing the end of the experiment, a loud pop echoes from across the room and glass shards scatter about the lab. An instant later, Aaron's screaming in my face. "I can't believe you actually tried to kill me, Joe. What the hell's wrong with you?"

"I had nothing to do with that, Aaron." *Damn it!* How do I always end up in middle of this shit? Seriously, what are the odds he'd actually blow something up today?

"Like I believe that. Everybody here heard you threaten me," he yells loud enough to broadcasts it throughout the entire lab.

"You screwed up all by yourself," I yell back.

"No, you screwed up this time, Joe, and you're going to pay for it!" He hollers this in the direction of the instructor, ensuring he hears it.

There goes the rest of my afternoon.

"Why did you miss football practice?" Lance asks when he gets back to our room after dinner.

"Because Aaron accused me of sabotaging his chemistry experiment. The damn prick actually had the professor believing it."

"What happened?" Lance's expression is a mixture of worry and curiosity.

"Aaron picked up the wrong flask, a four-liter Erlenmeyer—"

"And since the four-liter Erlenmeyer flask is structurally incapable of withstanding extreme pressure, kaboom!" There's more than a hint of excitement in his voice. A moment later, he's serious again. "But it's all straightened out, right? You're off the hook?"

"Yeah." No thanks to the viper Medusa, who gladly ratted me out for the threats I said to Aaron. Good thing Anna and Trisha vouched for me.

"For drama on a different front, Joe, three worlds will be making a visual spectacle tonight, and the triple planetary alignment will be visible with the naked eye!"

"Hooray," I sarcastically mumble.

"Hey, here's a *solution* for you; instead of recklessly provoking Aaron, why don't you act sickeningly sweet to him? I bet it'd drive him cray-cray." Lance starts grinning from ear to ear, proud of his creative "solution" to my Aaron chemistry problem.

"I don't want him crazier than he is now." Wait a minute, that is a good idea. Maybe I've been going about this the wrong way. "Thanks, Lance."

Grabbing my jacket, I head over to Anna's room to test out this new tactic. Instead of fighting the attraction between us, I'll encourage it and hopefully use it to my advantage.

Chapter 15

It is not in the stars to hold our destiny, but in ourselves.
William Shakespeare

On October 6, 1995, humanity discovered a planet orbiting another star in our galaxy. The centuries-old quest for finding other worlds like Earth took a giant step forward. The name of this first planet is 51 Pegasi b.

Tatiana
Thursday evening, October 4

I'm lying on my bed, listening to music, when Joe knocks loudly, leans against the door jamb and smiles at me. I'm not sure if I'm glad or annoyed that Hannah left the door wide open again. She has a bad habit of doing that whenever she's in the room.

Joe holds his hand out toward me as if delivering a letter. "I got us two front-row tickets for tonight's special event starring three dazzling performers. Venus, the most brilliant, Jupiter, a distant second and Mars, trailing in third," he smugly states, enjoying his clever play on words.

I immediately recognize what he's telling me. A triple planetary alignment. These take place every year or two, but often they're too close to the sun to view or one of the planets is too faint to see. Triple alignments when all three planets shine brightly are rare. I forgot this was happening tonight.

"The show's about to begin. Come and get starstruck," Joe brazenly invites me, beckoning with his outstretched hand.

Hannah's instantly standing protectively near me, before I'm even sitting upright. She hasn't taken her eyes off of Joe since he appeared; she's more suspicious of his intentions than usual. More than likely, one of her friends filled her in about what happened in chemistry lab today. However, since Joe and I have been conversing in Russian, she doesn't know what's going on between us or what she should do about it.

Joe motions again for me to join him. Getting up off the bed, I pick up my coat and walk toward him. I'm glad about the prospect of spending more time with him and thrilled about the triple planetary alignment.

"You shouldn't go anywhere with him, Anna. He doesn't give a hoot about anyone except himself and he thinks the sun comes up just to hear him crow." Hannah firmly grabs me by the arm.

"Sun not up now, Han-nah. It eez nite out," I explain to her as I pry her fingers off of me.

"Where ... are ... you ... go ... ing?" She asks very slowly, emphasizing each syllable as if I can't understand her.

"Go ... ing ... du ... star ... show," I answer, just as slowly.

She turns to confront Joe. "Goodsell Observatory only has open houses on Fridays. It's late, and Anna has an early class tomorrow. Where are you taking her?"

Enough already, Hannah. I don't require any additional keepers. I've already got one more here than I want or need, which is none. I start zipping up my coat, but it gets caught.

"Goodsell Observatory has three historic telescopes," Joe explains to me in Russian as he pulls my zipper free. "A 16.2-inch John Brashear refractor, an 8.25-inch Alvan Clark & Sons refractor and a transit telescope originally used to set time in the Midwest."

I can't wait until Fridays! Excitedly, I move to join Joe, not caring if my jacket is zipped or not. With the speed of a jackrabbit, Hannah decisively wedges herself between us, going full upright on her tippy-toes in an attempt to glare directly into Joe's face.

"You haven't answered my question," she snipes at him. *Welcome to the world of Joe, Hannah. He never answers questions directly.*

Stepping around her, Joe mockingly summons me in English. "Come, gentle night; come, loving black-browed night, give me my *Anna,* and when I shall die, take her and cut her out in little stars, and she will make the face of heaven so fine that all the world will be in love with night and pay no worship to the garish sun."

"Quoting *Romeo and Juliet* is not answering my question." Hannah pivots with Joe's movements to keep her target front and center and starts poking him in the chest. "Making reference to cutting Anna up and turning her into stars is not silencing my alarm bells about you."

Uh-oh. How is he going to react to being jabbed like that?

Joe half smirks. "No, I suppose it wouldn't." Reaching past her, he grasps my hand. "Let's get out of here and go get starstruck."

Ypa! I do an internal happy dance as I leap around Hannah. No way am I passing up this golden opportunity, seeing a relatively rare astronomical event and getting private time with Joe. Fortuitously, Ms. Mizuikova departed for the night just five minutes before he arrived. She has a rented unit somewhere nearby where she goes to sleep when she's not busy interfering with my life.

I jubilantly bound all the way down the stairs, relishing the pleasurable feeling I'm getting from defying my father's strict new edict—until he has determined whether Joe is a danger to me, I am not to talk with him, be near him or go anywhere with him. *Do you realize how ridiculous that is, Father? We're lab partners.*

Once I'm outside, the cool, invigorating night air tickles my exposed skin with a hint of the exhilaration to come. We find a spot with an unrestricted view of the stars, sit side by side on the grass and gaze upward to search the crystal-clear sky. "There it is!" We exclaim in unison, pointing to the triple alignment.

No matter how many times I peer into the window of the universe, the indescribable beauty and sheer vastness of it amazes me. I picture myself sailing among the cosmos, and when I reach Alpha Centauri, I pause to listen for the celestial prelude … I can almost hear the singing. Then the voice from my dream, the old woman's voice on Mars, cuts in. *I saw past the galaxies into the infinity of darkness, and I saw the star thrower. I saw you.*

This time, instead of being put off by her ominous words, I'm curious. "What's a star thrower, Joe?" I don't know why I haven't looked this up.

" 'The Star Thrower' is part of a sixteen-page essay of the same name by Loren Eiseley, published in 1969 in The Unexpected Universe." Wow! I didn't expect such a detailed, factual answer. I was expecting some sort of whimsical reply. I'll need to read that article. "When did you become so fascinated with space, Anna?" Joe asks.

"Ever since I took my first breath." Inhaling deeply, I continue absorbing this exhilarating sense of cosmic wonder that I'm truly *feeling* for the first time, or at least since the start of my long hiatus from emotions. "When I was a little girl, not long after my mama died, I learned about a planet, 51 Pegasi b. It's the very first confirmed exoplanet that orbits a star similar to our sun."

"So?" Joe doesn't see the importance, or why it should have such an impact on me.

"So, it's the first solid proof that there are other planets like Earth out there, and it represents the reality of viable worlds beyond this one. Even though it's fifty light-years away and most people are oblivious of its existence, 51 Pegasi b forever altered humanity's outlook of the universe." It sure altered mine; it allowed me to envision a different future.

"51 Pegasi b isn't a habitable planet," he counters.

"That doesn't matter. It will always be special." Edging over, I nestle closer to Joe. If experiencing feelings are here to stay, I'm going to start thoroughly enjoying the good ones.

When our shoulders touch, Joe tenses, but doesn't move away. "I've been reading up on NASA's Kepler Mission, Anna. They have verified 715 new worlds, bringing the current count of planets outside our solar system to 1,794."

"I know. Isn't it wonderful? And there are billions more planets out there. Virtually every star in the sky has at least one planet around it." A jittery buzzing hums through our bodies, taking effect like a sugar high. Excitedly, I pounce on the easy lead-in he's given me to learn more about him. "When did you first become fascinated with space?"

"After I saw my first *Star Trek* movie."

Yolki-palki! Not him too! My father and brothers are ridiculously obsessed with *Star Trek*. Consequently, I've had long years of being subjected to their endless quotes, which my brothers often directed at me. I immediately change the subject. "Have you ever read about the cosmonauts' reactions, the ones who flew to the moon?"

"Sure. Many of them remarked that the most important thing they discovered was Earth. Exoplanets must not have been on their radar scope," Joe teases me.

"You totally bypassed the most important part, their descriptions." I emphatically point one hand at the moon and tilt his chin in that direction with the other. "Imagine yourself up there. Before you is an astonishing view, a striking blue planet Earth gleaming against the dark tapestry of space. Buzz Aldrin described it as a brilliant jewel in a black velvet sky." As I recite this, the here melts away and I'm there on the moon. It's spectacular! I've never imagined anything this vividly before. Is it the feelings causing this or Joe?

Something deeper starts poking at me, prodding at the edges of my awareness. "Do you sense that, Joe, an elusive tugging, coming from out there among the stars?" He gives me an almost imperceptible nod. Yet even though this strange connection lets him experience what I'm feeling, he doesn't understand. "There's an intangible sense of ... human interconnectedness. Looking back at Earth from the moon, we're all connected, there are no separations. Individual differences no longer matter."

"Individual differences do matter," Joe firmly contradicts.

"Not when we comprehend the full scale of the universe," I assert. "On Earth, our world appears massive and enduring to us, but this planet is infinitesimally small. If only we could grasp the connection ... we need to keep it safe."

"Keep what safe?"

Poof! Just like that, my wisp of insight is scattered to the stars, and only a faint lingering of greater purpose remains. My hands fall to my lap.

"Keep what safe?" he repeats.

"I'm not exactly sure. A wholeness, a oneness with space and place, past, present and future." Intently I concentrate, trying to regain my hold on it. Nope, it's gone.

"There's nothing out there controlling our fate, Anna."

He sounds so convincing, but I don't believe him. In my life, there are definitely things controlled by fate. As my exhilaration fades, I begin softly singing. *"Cold and starry skies, hidden deep in Joe's own eyes, shifting—"*

"Why did you use my name in that song?" He scuttles sideways to put some distance between us. The extra space doesn't matter. My emotional tank overflows from the sheer volume of intense grief pouring off of Joe. His unstable emotions with their erratic yo-yoing are throwing me disturbingly out of whack. In the last half hour, I've gone from gleeful anticipation to unfettered wonder to disconcerted confusion to wistful melancholy and now excruciating grief. It's exhausting. Why does Joe's presence have to affect me this way? And why does my singing affect Joe this way? He had the exact same reaction last time he heard me singing, and I don't like it. I don't like my earthly refuge being twisted into a conduit for misery.

"Why does my singing upset you so much, Joe? And don't change the subject to avoid answering me."

He gets a faraway look on his face. " 'The inexpressible depth of music, so easy to understand and yet so inexplicable ... it reproduces all the emotions of our innermost being.' Arthur Schopenhauer stated that quite well. Do you agree, Anna?"

"Yes!" Maybe Joe does understand what I was talking about. That fleeting tendril of connectedness reappears and nudges me once more. *What if music connects people?* I wonder the idea out loud. "What if the songs in our souls resonate throughout the entire universe? Perhaps, it's music that connects all life."

"Dao De Jing," he muses. "The teachings of the ancient Chinese philosopher Lao Tzu. Legend says that his birth occurred when his mother gazed upon a falling star." Simultaneously, we both search the night sky. No falling stars tonight. "According to traditional accounts, Lao Tzu was an archivist in the library of the Zhou Dynasty Court," Joe continues. "When forced into exile, he rode a water buffalo to the edge of the desert where a guard stopped him. Lao Tzu wasn't allowed to leave until he wrote down his teachings. He was considered one of the greatest thinkers of his time, Anna."

How does he remember all this stuff? No, I'm not going to ask, I don't need him going off on a tangent when he's on a subject I want to discuss. "What did Lao Tzu write about music?" I ask, hoping one of the greatest thinkers of all time can provide some insight.

"Over two thousand years ago, he wrote what you just said. Music in the soul can be heard by the universe."

That's quite a coincidence. A sudden breeze whisks through my hair, blows down my neck and sends a shiver running through me. Is it a warning or a jolt of excitement? "Tell me more about Dao De Jing."

"Nope." Joe determinedly shakes his head. "Explaining anything more to you about that would be totally counterproductive."

The gauntlet has been laid down. "I want to know." He continues shaking his head and lets out a disparaging guffaw. You'd think I'd asked for the moon. "Tell me!" I insist.

"All right, Anna, just don't get mad at me when you totally fail at it." He scoots over a little farther so he's out of arm's reach. "The central concept of Dao De Jing is wu wei, which is non-doing or doing nothing. We need to live in harmony with the natural evolution of the cosmos. It's all based on the premise that people have desires and free will, and when we act irregularly, we upset the natural balance."

I scrunch up my face in disgust. "That's it? Do nothing. How does that resolve anything? There's got to be more to it than that. Explain exactly how I regain my natural balance and be very specific." I may regret asking this, because I'm not the least bit interested in philosophy. It's too wishy-washy and has never been of any practical use to me. However, this Dao De Jing may be different.

"Think of a large rock dropped in a river," Joe says. *Okay, I can do that.* I picture a boulder falling into the water. "The river does not stop, Anna. It flows around the rock and resumes its natural course." I visualize the water curving around the boulder and going back on course. But it doesn't always do that. It depends on how large the boulder is, in which part of the river it falls and if there's other boulders nearby. This is why I don't like philosophy.

"Doing nothing is the way to seek the calm state of wu wei," Joe evenly restates.

"You're talking in circles," I yell at him and lean way over to spin circles in front of his face with my index finger. "Circles, circles, circles. You're not clarifying anything."

Joe somehow manages to remain composed amidst my mayhem. "You return to your natural balance by not forcing things, Anna, by flowing with the moment." Per his instructions, I change my hand movements to flowing waves, and Joe shoves my arm away. "As you've just demonstrated, Anna, *wu wei* is something you have no possibility of achieving."

"Speak for yourself." *I have no difficulties keeping my emotions in check, when I'm not around you.* Enough wu wei. Back to what I really want to know. "Why does my singing upset you so much?" I slide right next to Joe and tilt my head so our faces are a few centimeters apart. As anticipated, a wonderfully intoxicating attraction thrums between us. Hah, this will *wu* him into confiding with me. *Uh-oh.* The lure is becoming so irresistible that I lean in even closer. Instantaneously, Joe pulls away. "We need to go. I can't be out past eleven on weeknights," he uncomfortably stammers.

"You have a curfew?" I'd better not let my father get wind of that idea, or I'll have one too. Although, in a manner of speaking, I already have a loosely defined curfew, but it has a huge loophole in it. My father decreed that unless Ms. Mizuikova is with me I am not to go out in the evening by myself. And technically speaking, I'm not by myself, I'm with Joe.

"We *star* athletes need our rest." Chuckling, he circles and waves his hand up at the sky, bemusedly poking fun at me. Even though I know he's purposely distracting me to avoid my question about singing, the temptation is too great. "Why do you have a curfew?"

"Actually, it's more of a guideline. The coaching staff is concerned I'll do something that'll get me expelled." He jumps up before anything too intimate can happen between us. I try to intercept his thoughts, hoping to figure out the best way to pry deeper without upsetting him. Unfortunately, the link between us doesn't work that way. I continue my mind-reading endeavor, and a totally beguilingly grin lights up Joe's face, half dazing me. His invitation from earlier in the evening pops into my head. *Let's get out of here and go get star struck.* "Huh, I just got star-struck," I comment.

His smile brightens, revealing an attractive dimple in his right cheek. "Elena Serova is slated to blast off soon in a Russian spacecraft for the International Space Station," he says, offering me his hand and effortlessly lifting me to my feet. "She'll be the first Russian female cosmonaut to live aboard the station and only the fourth to fly in space.

"Yes, I know," I reply as we trod back toward my dorm.

"It's been nearly fifty years since Valentina Tereshkova became the first woman in space. She made forty-eight orbits in seventy-one hours. That was more time up there than the total of all the US astronauts to that date."

"I know that, too," I irritably reply. The mention of Valentina's name upsets me. Why does everyone think she's so great? She's nothing special. I hated it when my father would hold her up as a role model for me. "What's your point, Joe?"

"She received the Order of Lenin, was honored with the United Nations Gold Medal of Peace and was given the title Hero of the Soviet Union. She was a pioneer for all women interested in going to space," he smugly answers.

"She's not a hero." Why does everyone think that? "Valentina was a civilian assembly worker with no military experience who quit school at the age of sixteen and never went back into space again because she was petrified the entire time she was up there. She had no business being on the Vostok 6 spacecraft in the first place!" Definitively, I smash Joe's assertion of her greatness, daring him to challenge me.

"She was an expert parachute jumper," Joe argues. "And after her spaceflight, she studied at the Zhukovsky Air Force engineering Academy, graduating with distinction as a cosmonaut engineer. Later, she earned a doctorate in engineering and went on to became the only female general officer in both the Soviet and the Russian armed forces. I would say those combined achievements qualified her to be recognized as a Hero of the Soviet Union."

Why won't he quit? I know he feels how angry I'm getting. "Do you know what her radio call name was? Chaika. The most obnoxious, annoying, irritating bird known to humankind!" *Chaika! Chaika! Chaika!* The taunting chant from my childhood fills my head. Classmates in my younger years would tease me with this word. Back then, it didn't bother me, I just ignored it ... but now!

"Seagull," Joe chirps at me in English.

"Shut up!" I explode.

"Seems like you have a chip on your shoulder, Anna, not the sort of thing a person who achieves wu wei would have," he cunningly points out.

"That's what you've been trying to do all this time? Prove I'm incapable of achieving wu wei? Well, congratulations, you win." Small tear droplets gather at the edges of my eyes. Horrified that he could manipulate me so, I race ahead into my dorm.

"I'm sorry, Anna. I didn't mean to upset you that much," he sheepishly apologizes as he follows me up the stairs. "Seagulls have a positive side too. They're highly intelligent and fly freely wherever they want. No one is their master."

I'm on the verge of crying as I rush into my room. It's not until I spot Hannah with her earbuds in, lip-syncing and dancing in place, that my rational brain snaps into place and I realize just how skillfully Joe played me. He wasn't trying to prove any point; he was purposely inciting me in order to get me off the topic of music and its link to the soul. Well, I'm not going to be outflanked that easily.

When he finishes creeping through the door frame, rather ashamed of what he did, I spring into action, pounce on him while he's still vulnerable and launch my attack. "What's your favorite song, Joe?" *What?* Why did I ask that question? It's not at all what I want to know.

"I need to go, Anna." He abruptly backpedals and bumps into a metal cage containing a brightly colored, stuffed parrot. Hannah keeps moving this ridiculous stumbling block around our room, trying to find it's ideal "roosting ground." I never know where it's going to turn up next. This time, I'm grateful it's located where it is and silently thank her, because she couldn't have placed it in a better spot to foil Joe's escape. I glance over at her and notice she's keeping an eye on us, but mostly, she's immersed in whatever music she's listening to.

"Don't be a rock in the river, Joe. Use some of that wu wei you've been throwing at me and flow with the moment." I attempt to keep a sense of hurt feelings in my voice, and amazingly, my feeble plea works. Not only does he stay, but he grants me a brief rendition.

"I'm a seagull, yet I'm all right. I fly where I want, to get out of sight."

"Absolutely fascinating." I'm amused by his improvised lyrics and thrilled by his rich, mellow tone. It's a travesty that he avoids music so; he has such a beautiful voice.

Joe graces me with another snippet of song. *"She's a hard-coiled, crazy, wild, wacky viper hunter, sure looks fleek to me. Well, she chased down Medusa, lit up a torch, trapped her in a corner and was ready to scorch."*

Listening to him rhythmically chant out these humorous stanzas, I get so giddy I actually break into laughter. It's glorious. I don't remember the last time I laughed. Suddenly, the jocularity comes to a crashing halt. Who appears at the door, but the viper herself. Looming in the doorway, Ms. Mizuikova fills the air with her malevolent venom, ricocheting it off the walls to inject it into Joe and me. What in the heck is she doing here this late in the evening? She's never done this before.

"Hey, Medusa. Did you like my song?" Joe heckles her, switching to English. "Are you out and about searching for your offspring? Perhaps Amphisbaena, the horned dragon creature with a snake-headed tail?"

Why must he always play with dynamite in a forest fire? He's going to ruin everything by recklessly provoking her. My curfew will be rewritten for sure. I didn't win the most recent battle with my father on either of two fronts, I wasn't able to clear up his suspicions about Joe's identity or convince him of Ms. Mizuikova incompetence, so I can't have her getting additional justifications for her actions. I rapidly lead her into the hallway and firmly shut the door behind us.

Slam the hatch, Lock out the demon! Keep the dangers way at baaay! The jolting lyrics pound at me, disrupting my thoughts. Not the ditty I need going through my head right now.

"Don't you understand how dangerous it is to be spending time with that boy?" Ms. Mizuikova's words blast through. "I thought your father made that perfectly clear to you!"

Slam that hatch, Lock out the demon! Keep the dangers way at baaay! The words keep pounding into me, making me wonder which side of the hatch the demon is residing in. "Why are you here My father made it perfectly clear that I'm allowed privacy," I snap right back at her.

She sternly glowers at me with her snobbish superiority, then she shoves a nine-by-eleven manila envelope in my face with a quick thrust of her arm. It's postmarked from Vermont and addressed to Anna Alkaeva, c/o Ms. Miesha Mizuikova, PO Box 1514, Eau Claire, Wisconsin. *T'fu.* Another of my father's security barriers. Apparently, he set up a drop point without any direct link to me as an emergency means of communication, *without* informing me of its existence. But who would be sending something to me this way and why is it postmarked from Vermont?

Aha! The address is my mama's older sister, Elena. She settled there after marrying an American journalist. I haven't seen her since I was a little girl though, and when I turned sixteen and started publicly speaking out against the Russian regime, she stopped mailing me letters and music CDs. Our family hasn't had any contact with her since then.

I cautiously rip open the envelope while Ms. Mizuikova obnoxiously hovers over me. Since her primary purpose here is my protection and this envelope could pose a threat, she won't leave until she's seen what's inside. However, I have no intention of sharing my private affairs with her. Having Ms. Mizuikova trail after me like a bloodhound on the hunt was my father's idea, not mine. She's the result of a last-ditch deal between us so that I could attend school here. Little did I realize back then what a big mistake that was.

Peering into the manila envelope, I see a standard-size white envelope postmarked from Moscow. I pull it out and tear it open to find a single sheet of folded paper. Shielding its contents from Ms. Mizuikova, I unfold the letter and scan the short, handwritten note.

T,

Ever since you left, I've felt so alone.

Without you, the world is such a dismal place.

The days are excruciatingly long and the nights even longer.

I can't wait to see you again.

V

"Give me the letter," Ms. Mizuikova spits as she coils herself centimeters away.

"It's nothing, just a short note from a friend," I nonchalantly dismiss her. Inwardly, I'm shocked, but I can't let her know that.

"Hand it over." Her voice is overly loud for such a public place that has relatively thin walls. She's trying to intimidate me. *Good luck with that, Ms. Mizuikova. You're a rank amateur compared to my father.*

I gladly point out her careless mistake. "Having a shouting match in the hallway is an irresponsible and stupid thing to do, especially with Joe on the other side of this door. He does speak Russian, you know. Get out of here, now." *Really, Father, couldn't you have hired someone a little more professional?*

"I wasn't hired by *you,*" she hisses back.

Hold on. The post office would have closed a long time ago. "Why didn't you bring this envelope to me earlier?" Her eyes momentarily dart away, and she fidgets uncomfortably, not having a ready reply.

"You've already read it! You opened my private mail without my permission." She stretches her neck out defensively and stares at me with beady eyes.

She's never going to admit it, and I have no proof. I've lost this battle. Petulantly, I throw the letter at her. It flits to the floor before she can grab it, and I take a small pleasure in watching her awkwardly bend down to retrieve it. After feigning scrutiny of the note, she shoves it back at me. "Who is V, and how did he know where to reach you?" she demands, attempting to regain control of the conversation.

"Viktor Agapov. He's a friend from secondary school, and he has no idea where to reach me. That's why the letter wasn't delivered to me here on campus." I calculatingly feed her this misleading yet viable explanation.

"How did he learn about the PO box address?" she further

interrogates.

"You're in charge of security. You tell me." If she thinks I'm going to kowtow to her brutish questions, she's even more foolish than I thought.

"Answer me." Ms. Mizuikova's getting so upset she's overlooking the fact that this letter was originally mailed to my aunt in Vermont, who then forwarded it here. Knowing my father and his insistence on maintaining "levels of security" as well as I do, it's highly likely that she's unaware of my aunt's existence. I'm certainly not filling her in on that.

My attention shifts to the admittedly valid concerns Ms. Mizuikova raised. Who wrote this note and how did they know my aunt's address? I wish I could contact Aunt Elena and ask her, but with my father's protective measures on steroids that would end up setting off DEFCON 1 procedures, which might very well include my evacuation. I'll have to ferret this out another way.

"Anna! How did he know the PO box address?" Ms. Mizuikova irritatingly repeats.

To mislead her, I strategically hesitate, then tentatively answer. "I might have let it slip." *Ypa!* The fish goes after the bait, or more aptly, the snake goes after the rodent.

"Do you realize what an impetuous and headstrong blunder that was? Do not respond to this letter, and terminate your association with that boy Joseph immediately!" She's morphed into the form of a rattlesnake and is viciously shaking her tail so loudly she's alerting everyone within earshot. *Really, Father, you hired a ticking time bomb.*

"The letter poses no danger," I adamantly state. "Viktor doesn't know I'm at Carleton so there's no traceable connection to me. Unless *you* were careless. So slither away and leave me alone."

Somehow, through her haze of fury, she recognizes there's nothing more that can be gained tonight. Heaving an extra hiss, she retreats toward her burrow. I, too, release an additional breath, then lean against the wall for support. It took every ounce of my strength to bluff my way through that. The V on the letter is meant to stand for Viktor Paskhin. And he died over three years ago!

Chapter 16

Time and space are modes by which we think,
not conditions in which we live.
Albert Einstein

Joe
Just over three years ago

Dear Joe,

I am writing to tell you I will always love you, and I hope someday you will understand why I had to leave. You would have gone after my dad, and then you'd either be dead or in prison. I couldn't live with myself knowing I was the cause of that. You're meant for greater things.

We have different paths to travel now. For once in your life, listen to me! Don't come after me. We made our music together, and I will cherish that forever.

Love, Lina

Joe
Two and a half years ago

"Where have you been?" Beth screams at me. "I've gone half mad thinking you got yourself killed! When you take off on these wild-goose chases, can you stop for just a second to text me and let me know you're still breathing?"

My sister's eyes are red and swollen from crying, and it makes me hate myself for causing her so much pain, but I can't stop. I have to keep searching, because someday I'm going to find her.

I'm going to find Lina.

Joe
Thursday night, October 4

After Anna goes into the hallway with the viper, Hannah fixes her attention on me, bluntly sizing me up and down, trying to figure out what she just witnessed. Parking her hands on her hips, she dishes out her disapproval, somewhat like my head coach after I've disregarded the sent in play and called my own, only to end up losing yardage.

"I've been instructing Anna to stay away from you. But does she give any credence to what I say? No."

Why would Anna to listen to you? Somehow I bite my tongue and don't say that. "Nice bird you got there, Hannah." I point to the cage I almost tripped over. "Better keep an eye on it or the viper will swallow it whole."

She stomps her foot, or at least as much of a stomp as she can muster in slippers, then forcefully pokes me with her index finger. "You're lower than a snake's belly in a wagon rut, and Anna doesn't need the likes of *you* in her life."

Jeez! Her nails, which are currently painted bright yellow with purple swirls, cut into me. At least she's sporting Vikings' colors. With a critical eye, she looks me up and down again, then questioningly tilts her head, seemingly confused about something. "The only time I see Anna show any positive emotions is when she's with you. Tonight, when she actually chuckled, I was nearly knocked off my rocker. Why doesn't she want to interact with me or any of her other classmates?"

I'm not touching that loaded question.

Hannah furrows her eyebrows, which are currently colored sunflower yellow, and continues processing her thoughts out loud. "I've tried everything I can think of to get her to chat with me. Since that seems about as likely as pigs flying, I guess that leaves you. But mark my words, if you so much as cause her one iota of pain, I'll knock you into the middle of next week looking both ways for Sundays." To reinforce her threat, she gives me another finger-stab to the chest.

Ow! For someone of such small size, she sure packs a punch. I don't complain though, because I probably deserve it for being so rude to her last spring when she insistently came on to me and wouldn't quit. Evidently, she's still holding a grudge, more than likely from the last thing I said to her. *Go to hell. They need talented people there with your skills.*

"Do not respond to this letter." Ms. Mizuikova's booming Russian words barrel through the closed door. "And terminate your reckless association with that boy Joseph immediately!"

Hannah's scowl deepens and her pointed index finger of power prods the air in the direction of Medusa. "That women rubs me the wrong way, constantly hovering about Anna like a bee over honey, except with the personality of a hornet. The only time she ever gives me any regard is when she tells me to get out of her way. Why does Anna put up with her?"

I'd like to know the answer to that too.

"Oh, fuff! Why I am talking to you about any of this?" She waltzes over to her desk and sits down.

I begin looking around the room, noting the stark differences between the two sides. Hannah's is covered with flamboyant decorations and photos; Anna's has no personal effects at all. Then something catches my eye, a Global Express box precariously perched on her trash can. Walking closer, I see it was shipped from Hammersmith, England. As I'm about to pick it up, Anna enters the room with a mixture of dread and fear trailing her. Blowing by me, her mind obviously on other things, she throws two envelopes on the far side of her bed. There's a large manila envelope and a smaller standard-size envelope. I surreptitiously glance over at them. *Tatiana ... 20 Wildberry ... Proctor.* That's all I can make out since the larger envelope is flipped upside down and overlaying the smaller one.

"Don't you have a curfew to make?" Anna lashes out at me.

What the heck is in those envelopes? She's obviously not going to fill me in on it. On the plus side, I've gained some concrete information directly related to her. Anxious to see where it leads, I follow her cue and take off. My brain shifts into overdrive, trying to decipher any meaning from those few words, but my memory can't find anything helpful. I'll require the ubiquitous, informative world of the internet.

Upon entering my room, I spot my sister Beth perched on the edge of my bed. Her coat is off and her cheeks are flushed. *Damn it!* Now is not a good time for a sibling visit, especially hers. We've always been closely tuned into each other, and she'll definitely notice the change in me. Then she'll dig her way into the middle of it to find out the reason—Anna's arrival in my life. I don't want that.

"Where have you been?" she promptly asks. "I've been waiting here for you since nine p.m."

"Out stargazing, Beth."

She sends me a disapproving look. "I called you numerous times. Funny thing, though, after your roommate let me in and I tried calling you yet again, I heard ringing close by. Guess where it was coming from? *Your* phone, right here in this room. Imagine that."

"Must have forgot it," I offhandedly remark, averting my gaze.

"You agreed to carry it with you when you're off tromping around." Letting out a disgruntled groan, she moves on. "I came down from Moorhead with the debate team for a competition at Bethel University in Saint Paul. Since I was so close, I thought I'd drive over to see you. Luckily, I managed to borrow a car from a friend."

Knowing my sister as well as I do, she didn't drive all the way here just to say "hi." Under the guise of a casual visit, there looms a deeper motive. "What do you want, Beth?"

"Ellie's brother Russ asked me to tell you thanks. He's attending Minnesota State University in Moorhead now."

She's still ducking around whatever it is that has her worried about me. "It's *Lina*, not Ellie," I remind her. "And you didn't make the trip just to tell me that."

"No, I didn't." She stares at her feet and absentmindedly twists one ankle over the other, wanting to hide her concern. Neither one of us is good at keeping a straight face, and especially not around each other. "I came because I always worry about you on September twenty-second ... I wanted to see you sooner, but it didn't work out. So, how are you, Joe?"

"Fine." She's close to succeeding in making me feel guilty for wishing she would leave, almost. But I don't want to spend the rest of the evening rehashing September twenty-second.

"You've never been fine, not since Mom died and certainly not since Lina disappeared." Beth's brilliant blue eyes, the exact same shade as mine, drill into me. "How are you *really*?" She's continuing her standard M.O. of leading with an innocuous greeting, telling me she's worried about me, emphasizing why she's worried about me and then prodding me to "work" through it. *That's not on this month's agenda, Beth.*

"I'm fine enough that I don't need coddling from my baby sister. I can take care of myself." Since she's only four minutes younger than me, this ploy sometimes works to get her off my back.

"Did you manage to stay out of jail this year?"

I know where she's going with this, the same place she always goes, lecturing me to stop obsessing over Lina. Since she's not letting up, best to get it over with. "Say what you need to, Beth."

She glances out the window, even though it's too dark to see anything from inside the lighted room. "There was an article in the *Star Tribune* yesterday, and since you normally scour the news daily, I automatically assumed you'd read it. After realizing you hadn't, I don't want to bring it to your attention."

"What article?" I'm hyperalert now.

She smooths out some wrinkles on my bedcover then turns back to face me. "A write-up about a young woman murdered in Minneapolis on Tuesday. Her description matches Lina's."

A knot of angst grips my gut, and I search the internet. It doesn't take long to find the article she's referring to.

Fatal Shooting in Northeast Minneapolis. *Police are investigating the death of a woman found in an abandoned car outside an apartment complex early Tuesday morning. No arrests have been made yet. The woman was in her early twenties, 5'8", 135 pounds, blue eyes, auburn shoulder-length hair and a small tattoo on her left ankle. Anyone with knowledge of the woman's identity is asked to contact the police department.*

"It's not Lina," I state with unequivocal certainty. Beth gives me a doubtful look. "It's not her," I insist.

She observes me more closely. "You're different now. Last time I saw you, that article would've sent you charging off like a bull in a China shop to investigate the matter for yourself."

"I'm not always bullheaded, Beth."

Disregarding my comment, she continues her intense perusal. "Are you involved with another girl?" There's a hopeful expression on her face.

"Not just any girl, a Russian girl," Lance chimes in as he enters the room and sets down his backpack. His timing is impeccable, in the wrong sort of way. This situation is quickly moving from bad to worse.

"Return from whence you came, Sir Lancelot." I ball up my jacket and throw it at him. I don't want them double teaming me, and I don't want Lance finding out about Lina and feeling pity for me.

There's a multitude of questions on his face as he runs through his options. He didn't miss the fact that Beth thinks I'm in a relationship with Anna. After a bit of indecision, he opts not to quiz me in front of her. "I guess I have somewhere I need to go. Wonderful to see you again, Elizabeth." He winks at her, then turns to me. "You've got five minutes, Joe."

Beth bounds up off the bed and grabs both my shoulders, excitedly pleading, "Who is she? What's she like?" Of all my siblings, Beth and I are the most alike. We have the same thick, wavy hair, the same intense blue eyes, the same dimple when we smile and the same annoyingly irritating stubborn streak. Telling her it's none of her business won't work. She'll just keep pestering everyone I know until she finds out.

"I'm helping someone with chemistry, that's all." It's nothing more, I firmly think to myself, but my heart won't let me believe this little lie.

"I wish it was more," she whispers. "You need to live your life again, Joe. None of it was your fault."

Why do we always end up talking about this? Brushing her arms away, I reach over and snatch up a box of granola bars from my desk. *What?* There's only one left.

"I had to do something while I was waiting for you," Beth complains.

Opening the mini fridge, I grab the chocolate milk carton; it's also noticeably emptier. After chugging down what's left of it, I throw it in the trash, landing it squarely on top of all the granola bar wrappers.

"I got a little thirsty after eating all those bars," she defends herself. Picking up her purse, Beth rifles through it, pulls out an Almond Joy and offers it to me. I take it from her and grumble thanks. Almond Joys are my favorite.

"I know you don't want to hear it, Joe, but you need to let Lina go. It's what she would want."

Why does everyone else seem to believe they know what Lina wants? I stop mid chew to set her straight. "You don't know a damn thing about what Lina wants!"

"I think I do, Joe. A week after she disappeared, I searched through your things and found the letter she wrote you."

"You went through my stuff?" I explode.

Beth meets my angry glare. "I didn't know what else to do. You were in total self-destruct mode, and everyone and everything in your wake was being destroyed, especially me." Her voice is cracking, and I'm suddenly conscious of how much my actions affected her. Growing up, we weren't inseparable as some twins were. She was the good twin, I was the bad twin, but there's always been a bond between us.

Beth retreats to the far edge of my bed and busies herself by gathering up her things. I sit next to her and she stills. We share a silent acknowledgment; *I care about you.* Regrettably, she can't stay quiet for long. "You know Lina better than anyone, Joe. Do you honestly believe she'd come back now to be with you?"

I can't think about it that way.

"Well?" she persists.

I won't let Lina go. I can't let Lina go. My memory won't allow me to do that.

Beth considers more ways to convince me. Thankfully, the uncompromising look on her face soon eases. "It's late. I need to get back to Saint Paul. Take care of yourself, Joe." We rise in unison and she gives me a tight hug. "Don't worry about walking me to the car, I parked nearby."

As much as I'd like to take her up on that, it's not happening. Picking up my coat from the floor, I shake off a few leaves that were caught in the hood and put it on. It's a somber trip to the parking lot. As I watch her unlock the door, I'm reminded of the time I convinced Lina to "borrow" her father's car at two a.m. and meet me by the lake. I never in my wildest dreams thought she would do it.

"We'll get through this, Joe." Beth gives me one more firm embrace, throws her purse on the passenger seat and jumps in. The car pulls out of the lot with her hand sticking out the window for a last wave goodbye.

The second I'm back in my room, I'm on the computer analyzing the few pieces of info I gleaned from that envelope and the box in Anna's room. Lance is too busy doing his own thing on his computer to razz me. First off, I search for any towns in England named Proctor. None. Next, I search for US towns named Proctor. Twelve hits come up. Narrowing the parameters by including *20 Wildberry,* I'm left with one match, Proctor, Vermont. Employing Zillow and following up with a look into public records reveals a single-family, three-bedroom, two-bath, 1,498-square-foot home on a 6,098-square-foot lot owned by Elena Proctor.

Who is Elena Proctor? She's the band director for Rutland Middle School and plays in the Vermont Symphony Orchestra. She's lived at this address for the past twenty-three years and was married to Robert Proctor, an American journalist who died eight years ago. They didn't have any children. The maiden name on her marriage certificate is Alliluyeva. That's noteworthy. I doubt her maiden name being Russian is a coincidence. Odds are she's related to Anna.

Typing *Elena Alliluyeva* into google, I get one hit—Svetlana

Alliluyeva, Joseph Stalin's daughter who died in Wisconsin in 2011. That's a bizarre fluke. Could Anna be a distant relative of Joseph Stalin? It might explain some things.

I research the local area. Proctor, Vermont is a very small town founded in 1886. It's home to Wilson Castle and is the former marble capital of the world. Both the United States Supreme Court Building and the Jefferson Memorial were constructed with Vermont marble. Cool fact, but irrelevant.

Next, Hammersmith, England. How does this city tie into the picture? It's a commercial center of London located on the north bank of the River Thames. Several multinational companies are located there and it's a key transport hub on the main road from central London to Heathrow Airport. This is a clearer connection to Anna. London, or more aptly Londongrad, is a haven for wealthy Russians, and Anna and her family are definitely wealthy Russians. I know this because even though Anna tries to appear casually dressed in clothes that are utilitarian and practical, she occasionally slips up. One day she wore an extremely expensive item, or as my sister Beth would say, "haute couture." It was a mid-length, Burberry trench coat, Britain's best-known, high-fashion clothing brand. I recognized it right away since all through high school Beth would shove fashion magazines in front of my face and point out Burberry outfits I should buy for her when I "make it big" in the NFL. This coat was one of them.

I scan more articles about Hammersmith and something catches my eye. Composer Gustav Holst. He lived across the river from Hammersmith for nearly forty years, during which time he composed *The Planets, Op. 32*. Yes, logically, I realize it's highly improbably that it's relevant, but Anna's so drawn to astronomy and music that I'm compelled to read about it.

Gustav Holst named each movement of *The Planets, Op. 32* after a planet in our solar system and its corresponding astrological character to convey the emotional influence of planets on the human soul. *Not again.* More references to music and its connection to the soul. I should quit reading this ... but I don't.

There are seven movements in *The Planets, Op. 32.* Mars, the Bringer of War, representing the God of conflict and bloodshed. Venus, the Bringer of Peace, representing the goddess of love and beauty. Mercury, the Winged Messenger, representing the messenger of the gods and rulership. Jupiter, the Bringer of Jollity, representing the ruler of the gods, their guardian and protector. Saturn, the Bringer of Old Age, representing the founder of civilizations, associated with limitations and restrictions. Uranus, the Magician, representing the personification of the heavens and the night sky, associated with genius. And lastly, Neptune, the Mystic, representing the god of the sea, associated with creativity and idealism, but also confusion and deception.

Against my better judgment, I use *sound* reasoning and listen to a full rendition of *The Planets, Op. 32.* Fifty-five minutes later, I'm left with an eerie feeling of unrest. The composition is both unnerving and simultaneously one of the most evocative pieces of music I've ever listened to. Long after it's ended, Holst's planetary messengers continue to orbit through my head—Mars, the harbinger of bloodshed, Jupiter, the guardian, Neptune, the master of deception and Venus, the goddess of love. What are they trying to tell me?

Pull it together, Joe. Facts. Stick to the facts. Okay, what can I conclude from the last bit of data I saw on the envelope, *Tatiana?* Could Anna's real name be Tatiana? I'll have to call her by that name when she's distracted and see if she reacts. Stretching my taut muscles, I start compiling what I've found out. There's Elena Proctor, living in Proctor and possibly acting as a proctor for Anna. *Strange.* I'll need to probe further into that relationship. In addition, there's someone in London sending Anna packages. Who's doing that and what are they sending? Too many unknowns.

I glance at the time; it's already two a.m. Lying on my bed, I close my eyes to rest, but my mind won't quit. *What are you hiding, Anna? What secrets are locked in the dark corners of your soul that no one is allowed to spy upon?* I'm suddenly overpowered by an intense urge to protect her.

Something's wrong. Something is very, very wrong.

Chapter 17

Tatiana
Four years ago

I lay my head on Viktor's shoulder and close my eyes, trying so hard to feel something, anything. Happiness, tenderness, joy, love. Concentrating with a fierce intensity, I attempt willing into existence, yet all I feel is the touch of his hands on my back, not a single emotion. Defeated, I push away.

"I wish I could break the chains that bind you, Tatiana." His eyes are glued to me.

"Must you be so melodramatic, Viktor?"

"Why won't you ever show me any affection?" He reaches out for me.

"I can't do that. We've been over and over this." I shove him away.

"Can't or won't? Let me into your life, Tatiana." His features are molding into that grimace he's taken to giving me when I frustrate him.

"Nothing has changed, Viktor. I can never be that person to you."

Tatiana
Thursday night, October 4

With the arrival of this letter, imminent danger is again knocking at my door. As much as I hate to admit it, Ms. Mizuikova is right, I shouldn't have allowed myself to get so close to Joe. It's reckless and dangerous.

As soon as I get him out of my room, I sit down and begin analyzing the letter. The V is meant to stand for Viktor Paskhin, the only person I ever attempted to have a relationship with. Not surprisingly, I was incapable of arousing any feelings toward him. We did, however, form a unique alliance of sorts. Then on September twenty-second, nine months after his conscription in the Russian military, he was killed. Therefore, it's incomprehensible that this letter, dated five weeks ago, was written by him.

I examine it more closely, inspecting the style and cadence. The writing is consistent with Viktor's. Could he still be alive? *Don't be ridiculous, Tatiana.* But who would know Viktor well enough to forge this? Like me, he was a loner. Viktor had no close friends.

"What did Ms. Mizuikova want? Why are you so upset?" Hannah's high-pitched voice shatters my concentration.

Must she screech that way all the time? I fold up the notepaper, pull out the lockbox from the locked drawer in my desk and stuff it inside, along with the envelopes. After relocking everything, I lie down on my bed and turn to face the wall. For a long time, my thoughts spin over and over the letter. Who sent it and why?

The lights go off, and before long I hear the now familiar rhythm of Hannah's breathing when she's asleep. I slip out of bed, as quiet as Hannah's stuffed parrot, and retrieve the envelopes. Maybe there's a clue on one of them I missed. Needing additional light, I activate the flashlight feature on my cell phone and start scanning the smaller envelope that was postmarked from Moscow and mailed to my aunt. On the bottom left corner, there's a small, oily stain which I sniff. There's a faint scent of gun cleaner. While that's not startling, it bothers me more than I want to admit.

The larger envelope was postmarked from Proctor, Vermont and mailed to the PO box in Eau Claire. This handwriting appears to be my aunt's. There's nothing else noteworthy about it, so I return to scrutinizing the smaller one. What am I missing? Focusing so hard that I feel my pulse beating in my temples, I intently glare at it. Could Viktor still be alive?

Agh. While having a message sent to me from a *dead* man is disturbing, what really concerns me is that this letter found me at all. The only person in this world who knows where I am is my father, and he would never reveal my location. Yet somehow this letter managed to reach me? How did this *Viktor* person know about my aunt and where she lives?

I glance at the time. It's two a.m. and I'm totally exhausted. Add in the additional stress of worrying that when my father learns of this he'll insist I move again, and I reach total burnout. Wearily, I rise to stow the envelopes away before I end up falling asleep with them in plain sight. Too tired to undress, I collapse back on my bed. All of a sudden, an aching anxiety overtakes me.

Something's wrong. Something is very, very wrong.

II

Part Two

SECOND MOVEMENT IN THE COSMIC SYMPHONY OF LIFE

Saturn

The harbinger of limitations and restrictions

Chapter 18

The ruthless world of Russian politics. Since 2001, Vladimir Putin has been crushing the most powerful oligarchs in order to build his authoritarian regime. To escape jail, many of them fled to England. In 2006, these exiled Russian oligarchs and their associates began dying under mysterious circumstances.

On November twenty-third, 2006, in London England, Alexander Litvinenko, a close friend of Russian oligarch Boris Berezovsky, died after meeting with FSB agents. His tea had been laced with radioactive polonium-210. Before his death, Litvinenko had been working with British intelligence to investigate links between senior Kremlin officials and Russian organized crime.

On February twelfth, 2008, in Surrey, England, Arkady "Badri" Patarkatsishvili died unexpectedly of a presumed heart attack. Patarkatsishvili, an exiled oligarch, was a very close friend and business partner of Boris Berezovsky. Weeks before his death, he stated, "I have one hundred and twenty bodyguards, but I know that's not enough. I don't feel safe anywhere."

On November tenth, 2012 in Surrey, England, Alexander Perepilichny suddenly dropped dead while jogging outside his home. Traces of a rare and deadly plant poison, gelsemium, known to be used by Russian assassins was found in his stomach. Perepilichny, an exiled Russian oligarch and associate of Boris Berezovsky, sought refuge in Britain after supplying evidence against corrupt Russian officials.

Tatiana
Two years ago
Russia

As I exit the Shokoladnitsa café, a man jumps me from behind.

"Zatknís'!" He threatens to keep me quiet, then viciously twists my left arm up behind my back. Wrapping his other arm around my throat, he starts forcing me toward a side alley. I struggle and manage to get hold of a switchblade; I always carry at least two with me. Stabbing him deeply in the leg, I wrench myself free of his grasp. As I'm backpedaling away, one of my grandfather's associates runs past me.

Grabbing my attacker, he brutally cuts him in the side of the neck, severing his carotid artery. The man instantly falls to the ground and quickly bleeds out, taking his motives and his intentions with him.

Tatiana
Friday morning, October 5

During the brief lapses of sleep I managed to accrue, two competing problems tormented my subconscious, Viktor and Joe. Freaky images of them pursuing me through the streets of Moscow, the streets of London and finally the Carleton campus kept replaying on the inner screen of my eyelids.

Is there a connection between Viktor and Joe? I abruptly jerk up in bed. I need answers, and I'm going to drive up to Hibbing Secondary School today to get some. If I leave immediately, I can slip away before Ms. Mizuikova arrives to "watch over" me for the day.

Leaving the lights off, I quickly dress, grab the key fob for my Prius and slip out the door. Mercifully, Hannah didn't wake up. The less she knows, the better. I hurry down the stairs with my brain zipping through multiple scenarios for my future, none of which have positive outcomes. Another early-morning call from my father added to my internal mayhem. He listed more discrepancies with what he asserts is Joe's alias, then ended the conversation with a very stern warning to keep away from him *and* let Ms. Mizuikova do her job.

Humph! She wasted no time informing my father that Joe was in my dorm room last night. I can't believe that serpent actually duped my father into believing that her "legitimate" security concerns trumped my valid objections to her invasion of my privacy. At least I had one thing going for me, my father was distracted enough by other urgent matters that I was able to allay his concerns about the letter, for now. But that only happened because he has no idea who Viktor Paskhin is, or I should say, was. I never mentioned Viktor to anyone. Our relationship was a clandestine one, and I managed to keep it that way because back then my father's ridiculously over-the-top security measures hadn't been implemented; my public, political criticisms were still relatively tame, and the attack outside the Shokoladnitsa café hadn't happened yet. By the time it did, Viktor was already dead. That's why telling Ms. Mizuikova that the V stood for Viktor Agapov, a different classmate of mine, didn't raise any alarm bells. Evacuation procedures have been avoided, however, all that will instantly change if my father starts snooping further into this.

I'm weary and wary, and my day's just starting. As I exit the building, a rush of cold air assaults me. The wind's blowing viciously and a hard frost covers the ground. I'm three steps away from the door when Joe emerges out of the gloom, bringing with him an even deeper sense of freezing foreboding.

How did he find me here just now? The breeze picks up and whips around even harder. *Ohhh.* I'm so bone-chillingly tired of all the holod sobachii, dog's freezing cold, that invades my world. I want to scream out, *Make it stop, make it stop!* I want to turn back the clock to when life was different and I could climb into my mama's lap to bury my head in her warm bosom. Alas, wishes don't work that way. There's no fairy godmother, no Prince Charming and no guardian angel waiting to save me from myself. There's only one thing in this world I can count on, and sadly it's the fundamental truth professed by one of the most brutal dictators to walk this earth. "I believe in one thing only, the power of human will."

"Iosif Vissarionovich Dzhugashvili, also known as Joseph Stalin, man of steel, 1879-1953," Joe states.

"What?"

" 'I believe in one thing only, the power of human will.' " Stalin said that. He was also known for saying, 'Die, but do not retreat.' Red Army troops were forced to fight to the death. If they resisted, they were executed."

Misty puffs of breath ominously swirl between us in the cool, mourning air. *Yes, brain, I'm aware I substituted the word* mourning *for* morning. A worrying itch scratches at the edges of my dread, and I stop walking. "How did you know I was thinking about that quote?"

"I didn't, Anna." He stands centimeters away from me. That's unusual, normally Joe likes extra space between us.

"Then why did you mention Stalin just now?"

"Because you did. You were thinking out loud."

Just great. Another issue I have to be concerned about. "What else did I say?" I hope I didn't speak Viktor's name.

"Nothing."

The gusts of air continue wickedly attacking us, heavy with things left unsaid as another fear shudders through me. How many other things has he heard me blat out?

"I feel that way too sometimes," Joe softly admits.

"Like you're about to fall off a cliff?" I suck in a gasp of amazement. "How are you reading my mine—I mean, my mind?" That cursed Mir diamond mine. No matter how many times I bury it in my memory, it keeps drilling its way back, just like it does for my grandpapa.

Joe's eyes widen with surprise at the mention of the word "mine." It's like he's having a déjà vu moment. "No, I feel like the world's a bitter, cold place," he quickly clarifies. "In Verkhoyansk, Russia, temperatures drop to negative 93.6 degrees Fahrenheit, so cold that the moisture from your breath turns into droplets of ice. If you sip a hot drink after coming in from that frigid temperature, your teeth will crack."

Where is his mind off to now? Well, I don't need any additional cold in my life. Anxious to get under way, I rush past him and double-time it to my car, but he soon catches up with me.

"Verkhoyansk is referred to as Stalin's Death Ring, Anna. It's where he sent his political exiles."

What is he implying? Unexpectedly, Joe leaps in front of me and I smack into him. Invisible sparks zap between us, muddling my mixed-up feelings even more.

"Have you ever heard of tardigrades, Anna? They're tiny creatures that sustain life in extreme conditions by suspending their metabolism. That's how they survived outer-space temperatures of minus four hundred and twenty three degrees Fahrenheit."

"Tatty, my pet water bear," I reverently whisper. I haven't thought of her in years.

"Yes, water bears," Joe reiterates. "You need to understand they only survive the severe cold by replacing ninety-nine percent of the water in their bodies with a synthesized sugar. Otherwise, the formation of ice crystals would kill them ... you can't do that, Anna."

Wow! No one could wade through all his idiosyncratic, confoundingly connected collection of facts and make sense out of it. "I don't have time for your riddles, Joe." Leaning left, I jump right, wanting to hop around him. He easily counters my move and continues blocking my way. *Blin.* This trick looks so easy when I watch football players do it.

"The Water bear's ability to withstand extreme cold was discovered on a joint US-Russian mission, the Foton-M3," he persists.

I'm not going to get lured into discussing space travel with him, not when I have a dead man writing letters to me. Shoving him as hard as I can, I manage to dart by and resume my beeline toward the car.

"Fotons are Russia's unmanned recoverable capsules used to conduct scientific experiments. Foton-M3 was the eleventh mission," he calls out while relentlessly pursuing me.

"No, Joe, it was the twelfth mission," I automatically correct. *What now?* He already knew it was the twelfth mission. I can tell by the sense of satisfaction flowing out of him.

"I used to dream I could do what a water bear does, Anna, suspend my life and wake up centuries later to a different world."

That's enough. Does he really expect me to believe he dreamed about water bears? I square off to confront him. "Spit it out, what are trying to tell me?"

"Where are you going?"

I should have known I wouldn't get a straight answer. The remainder of the way to the parking lot, Joe trails a few steps behind me. At least, he finally shut up.

When my car comes into sight, I start rushing toward it. In my haste, I trip over a crack in the pavement and do a full airplane, landing with my wheels up. Could the start to this morning get any worse?

Joe's instantaneously by my side. Scrambling away, I hurry to stand up without any help from him because I'm afraid of where my feelings will go if he touches me. I need to be clearheaded. Taking inventory of my injuries, I note my left palm is bleeding and my right elbow is sore. Other than that, I'm relatively unscathed.

As I walk, not run, toward the car, my gait feels funny. I look down and see something stuck to the bottom of my shoe. Reaching forward, I peel off what looks to be part of a newspaper page.

CARLETON VICTOR. The juxtaposed position of these printed names screams out at me. Are Joe and this Viktor person working together? Is the universe trying to warn me?

"What's wrong, Anna? Does the fact that the Carleton football team achieved another victory shock you that much?"

T'fu. My imagination is scurrying off and dragging me through back alleys. Pulling out my key fob, I carefully sprint the final distance to my car. As I'm opening the door, Joe advises me in a foreboding tone, "Be careful, Anna."

Why? Why do I need to be careful, Joe?

I hurriedly plop into the driver's seat, start the engine and gun it. Hibbing, Minnesota, here I come. With any luck, I'll uncover some much-needed *chut-chuts* of Joe's history.

Four and a half hours later, I park my car along the curb in front of the Hibbing School. It's quite a massive and imposing structure. I amble up the front steps while slowly sliding my hand over the cold, brass railing for support, wondering what I'm going to find here. Is Joe friend or foe?

Inside the foyer, I'm stunned by the exquisitely ornate plaster ceilings and mosaic tiled floors. The exorbitant riches spent on this educational structure baffle me. Maybe I'm not in the right place. I enter the main office and search for someone to help me. There's a well-dressed, middle-aged woman standing behind a waist-high counter with a welcoming smile on her face.

"Eez diz Hib-bing Secondary School?" I ask. It looks more like a Russian Opera House.

"This is the Hibbing *High* School," she responds. Seeing the confused look on my face and assuming it was because she said 'high school', she begins to tell me the location of a nearby elementary school. In Russia, there is primary schooling, general schooling, secondary schooling, and then if you pass the Unified State Examination, higher education.

"Haz diz been de high school for a long time?" I ask.

"Yes, quite a long time. Iron ore was discovered under the town of Hibbing around 1918, and the entire community was moved. The Oliver Iron Mining Company built this school to promote the relocation. It is on the National Register of Historic Places, and—"

"Du yu hav copi of Hib-bing yerbook?" I cut her off, not interested in the town's history and definitely not interested in a history of mining.

"Certainly," she replies. "There are copies of yearbooks in the library."

Guided by her directions and thick marble pillars, I wander the hallway and find the library. The entire far wall of the room is dominated by a roughly eighteen-meter oil painting displaying miners at work. *Why must I keep being reminded of mines?* I wish I could purge that damn place and its cursed diamonds from my life.

Making my way over to what appears to be an information desk, I ask where the yearbooks are. With an air of indifference, the attendant points toward a bookcase in back. It takes a bit of searching, someone had stuck the one I want on the shelf out of order, but I find the 2011 yearbook. Sliding it out, I take the book over to one of the round wooden tables and sit down with my prize. What will I discover? *Who are you, Joe?*

I excitedly flip through the photos of graduating students and meticulously scan for Joseph Carleton. I can't find him! Starting over from the beginning, I scrutinize every student portrait in the entire yearbook and discover photos of Benjamin Carleton and Natalie Carleton, but no Joe.

I must be overlooking something. *Oh, yeah.* Joe didn't start college right away, he worked on his father's farm for a year first. I return to the shelf and retrieve the 2010 yearbook to initiate my quest anew with an even fiercer determination. Starting from the very first page, I thoroughly inspect every word and picture. On the bottom right-hand corner of page twenty-one, I spot a portrait of Elizabeth Carleton. She has a strong resemblance to Joe. Turning to the next page with baited anticipation, I'm expecting to hit pay dirt.

There it is! Joseph Earl Carleton. But the name is written under a grayed-out rectangle. Another dead-end. Why is there no photo of him? School pictures are taken toward the start of the school year, not the end, when Joe was expelled, or at least that's the time frame when we take them in Russia. Unlike these, our portraits are not so boring and mundane with the same bland color behind each person. Every student gets to create their own backdrop. I drew pictures of cats and images of my favorite constellations on the chalkboard I was seated in front of.

Frustrated by my mounting failures, I despondently scan more pages, searching for any picture of a guy who remotely resembles Joe. When I reach the point where students last names start with E, my attention is captured by a photo of Ginny Engel. She seems vaguely familiar, a pale complexion with long black hair and riveting dark eyes.

I remember. I saw her recently on Carleton campus going into Joe's dorm. She stood out because she popped in and out of the entrance two times in quick succession before finally staying inside. I wonder if she was there to see Joe. Could she be the one he took the blame for in the cheating scam? I know he told me it was a guy, but maybe he said that to mislead me.

I stare her image, and after a few moments, my focus is oddly drawn to the next photo. Eleanor Engel. Her bright sapphire eyes, which are partially hidden under wavy auburn hair, look right into mine, shooting an unexplainable feeling of wistful nostalgia into me. Who is this? Does she have a relationship with Joe? Are Eleanor and Ginny sisters? They don't look related.

I scour the remaining student photos and discover one more Engel, tenth grader Russel Engel. Storing these names for future reference, I resume my mission of finding a photo of Joe and begin examining the student-activity sections. Front and center of the show choir group shot is Eleanor Engel. She looks different here, sadder and ...

A faint melody drifts in. I recognize it right away, *Music of the Night* from Phantom of the Opera. I've never heard this rendition though, it's more bittersweet and haunting. Curiously drawn to the song, I exit the library in search of the source. As I venture down the hall, I hear it more distinctly. There's a violin playing. Following the ghostly strains, I reach a set of double doors, but when I touch a handle, the music stops. Before I lose my nerve, I wrench open the door and take a few steps inside. A breathtakingly elegant auditorium with four massive, crystal chandeliers hanging from the ceiling greets me.

"Can I help you?" A man's voice comes from behind me.

"Ver doez music com from?" I hesitantly ask.

"That would be our magnificent Barton pipe organ that was constructed in Germany. It's one of only three ever made." He stands a little taller as he recites this. "Containing over nineteen hundred pipes, the organ can reproduce any orchestral instrument except the violin."

Except the violin? I could swear I heard a violin playing. A creepy shiver runs down my spine.

He points to the ceiling before I can question him on this. "Those crystal chandeliers are from Czechoslovakia and were made in a plant that was destroyed by German planes during World War II." His eyes get a faraway look, and he distractedly runs his fingers through his thin, graying hair.

Does everybody in this school have to paraphrase sections of textbooks when they talk? Is this where Joe learned that annoying habit?

"Our auditorium is the centerpiece of the school and was designed to replicate the historic Capitol Theatre in New York City. Would you like to hear more about it?" He resumes his tour-guide mode.

No ... but sort of.

Encouraged by my slight nod, he walks toward the stage and beckons me to follow. "Look up there." I track his extended hand to see grand arches and pillars framing an enormous stage. "That fly area holds backdrops that rise up ninety feet. Forty-five backdrops were specifically created for us. This stage is still used for community events. In fact, that's why you heard the music. We're organizing for our fall performance."

We continue down the aisle, and I glance left and right at the lines of upturned seats. We pass row L, row K and row J. One of the seat bottoms in this row is stuck down. It appears to be the only seat in the whole auditorium that's out of place. Gravitating towards it, I take a step sideways when a gut-wrenching pain sucker punches me, taking my breath away and forcing me to double over.

"Are you all right?" My impromptu guide races toward me.

"Needt eer," I manage to wring out. I'm sure I look four shades past the grave. Taking hold of my arm, he leads me out of the auditorium and out the front entrance of the school. Drained of energy, I lean back against the brick building and draw in deep gulps of fresh air. My gasping gradually subsides, but not the anxiety from that mysterious panic attack. Looking over, I notice the old man is just as rattled as I am.

"Are you sure you're all right?" he asks again, his voice raspy and shaky. I feel like I should be asking him this. "Do you want to go back in and sit down? I can get you some water," he nervously offers.

"No." No way am I going back into that building. Nodding goodbye, I clutch the brass railing and make my way down the stairs. By the time I'm in my car, only a remnant of that horrendous feeling remains, but that nasty little bit is firmly implanted inside of me.

Chapter 19

The universe is leading us somewhere,
and along the way, it is changing us.
Natalie Batalha

Tatiana
Friday, October 5

The drive back to Carleton gives me time to think. One, two, three, four times, I replay what just occurred in the auditorium. The violin music, the strange seat and the panicky feeling. Everything connected with Joe just keeps getting weirder and weirder.

Anyway, I gathered a little more information about him. Joe has a sister who is either a twin or very close in age and at least two younger siblings. His "twin" sister was extremely active in school activities. Class officer, debate team, speech, student council, Spanish club, and probably a few more I missed. Joe seemed to exist only for football.

Since I was so bizarrely interrupted, I wasn't able to finish searching through the yearbook for a recognizable photo of him. Looking ahead, I should be able to electronically access old editions of the Hibbing newspaper and check their Sports section. I'm done with thinking, for now.

Cranking up the volume on the sound system, I put in one of the CDs my aunt Elena gave me and blast some much-needed music for the remainder of the ride to Northfield. I encounter no major delays and arrive on campus in time for an early dinner, which I bring up to my room. With my hands full of food and drink, I carefully open the slightly ajar door, then slam it shut with the heel of my foot.

Hannah glances over. "Medusa was here looking for you. That woman has no people skills. She could make a bishop mad enough to kick in stained-glass windows."

Blin. Now I have Hannah referring to Ms. Mizuikova as Medusa, too. I'd better not slip up and call her that to her face. I don't need the serpent turning vindictive on me. "Vat did yu tell er?"

"Nothing," she replies with a touch of resentment. "I couldn't *tell* her anything, because I don't *know* anything. Where you were, who you were with, what time you left, what time you would get back, what ..." She's ticking off the list on her fingers as she recites it. I tune her out before she switches hand. In retaliation for me ignoring her, or maybe due to Medusa heckling her, or probably both, Hannah starts belting out a song.

I set down the Styrofoam container of beef stroganoff and the glass of coke on my desk, then clamp my hands over my ears. She stops singing and ends with the word "stars."

"Who rote dat song?"

"Bruno Mars," she replies.

Mars! I need to get that music loaded onto my iPod. Yes, I'm a little obsessed with Mars, but I'm not the only one. Many people in the US and Russia are too. The first time a spacecraft flew by a planet, it was Mars; the first time a satellite orbited another planet, it was Mars; the first time a spacecraft landed on a different planet, it was Mars; the first time we roved around the surface of another planet, it was Mars ... and the first time I set foot on another planet, it's going to be Mars.

Get your head back to Earth, Tatiana, you have issues here that need your attention. I settle into my chair, power up my computer and google *Hibbing MN newspaper*. The *Hibbing Daily Tribune* pops right up. Clicking on the website, I immediately hit another barrier. I'm unable to access more than a few editions without paying for a subscription. Can't do that. It would instantly trigger the jaws of my father's security apparatus.

Movement catches my eye. Hannah's pulling the door wide open and wedging a bright pink doorstop under it with her foot. "Keeping the door closed isn't going to stop that snake of a women from entering, Anna, it's only going to prevent my friends from spontaneously bopping in to say a quick "hi."

Bet if I closed and locked it, I could achieve both of those goals, Hannah.

"I do appreciate the intentions behind your method though. I don't know how you tolerate that woman." She suddenly throws off her slippers and starts fervently trying to remove the doorstop with her bare feet. Her fingernails must be freshly painted since she's not using her hands for this task. A moment later, she abandons the foot technique and scrambles with both hands to extract it. "On second thought, maybe there are some advantages to keeping the door closed."

Why the sudden reversal?

Joe appears and scoots past Hannah just before the doorstop is dislodged. "I wanted to make sure you got back safely and I brought you these," he awkwardly states in Russian, hesitantly approaches me and hands me a box.

I recognize it right away, Red October chocolates. Red, like the Red Planet. I smile at the happy happenstance. Opening the lid, I gaze upon the most popular Russian sweet and my personal favorite, Alyonkas. The small squares are individually enfolded in foil, each with a picture of a rosy-cheeked little girl wearing a flowered shawl. Even though they're all identical, I select a chocolate in the middle of the box and reverently unwrap it. Lifting the Alyonka to my nose, I luxuriate in the familiar, delectable smell. Popping it in my mouth whole, I delight in the scrumptious flavor. *Mmmmm.*

"Where did you find these?" I ask as my taste buds swim in chocolate heaven. Picking up another one, I sniff the intoxicating scent. Abruptly, an unnerving thought intrudes. "How did you know Alyonkas are my favorite?"

"I can't give up all my secrets," Joe jests.

"No, you don't give up any of them. Almost everything I've learned about you is through my father." The frustration from my mostly fruitless and overly dramatic trip up to Hibbing permeates my voice.

"You know more about me than anyone else at this college, Anna."

"Like what? That you have a dog you call Princess?" Angrily slamming the box of chocolates closed, I shove it back at him. The hurt look in his eyes and the corresponding emotional sting that accompanies it take me by surprise.

"What upset you so much last night, Anna?" Joe gently slides the chocolates back to me. "Medu—" He speedily stops himself, sensing my extreme annoyance at this reference. "The creature paid me a visit today. She was in fine form, spitting venom and rattling off accusations, thinking I had something to do with your mysterious morning foray. Where did you go?"

"I drove up to Hibbing to determine if you ever attended school there. Did you?" I outright dare him to refute it.

"What do you think?"

"I think you're hiding something from me. I discovered that a Joseph Carleton did indeed attend school there, but I couldn't confirm that *you* are that Joseph Carleton because I couldn't find a clear picture of him anywhere. It was conspicuously missing from the year book. You wouldn't know anything about that, would you?" Rapidly surmising he won't elaborate, I start listing the other family member names I remember. "I saw photos of Elizabeth Carleton, Benjamin Carleton and Natalie Carleton."

Joe doesn't react. He stands there, stolidly eyeing me.

Might as well throw out the additional names. "I also found a photo of Ginny Engel." Now, I'm beginning to elicit a response. "And right next to Ginny, was a photograph of Eleanor Engel."

Joe brusquely strides over to the window and looks out with an aggravated frown, futilely trying to hide the intensity of his emotions. His ever-present guilt is extreme now, but he's also furious, very furious. I have the unpleasant realization that I may have poked the tiger one too many times. "What do you really want from me, Anna?" he spits out through gritted teeth.

Kaboom! His question sends emotional shrapnel heedlessly ricocheting inside my body. *I want to know who you are. I want to understand what's happening to me. I want to be able to experience my feelings and only my feelings, and experience them without dangerous consequences. I want to understand you, Joe, and not be separated by a multitude of secrets, because you're the only person since my mama died that's ever made me feel whole and safe and special. I want you to hold me again and tell me everything is going to be all right.* Yet I can't say any of that, so I simply state, "I don't know."

Random images appear in my head, an organ, a mining pit, a violin, then an abandoned playground. Then two things I absolutely don't want joining the wild ride jump aboard, the letter I received from dead Viktor and the gut-wrenching feeling that ripped through me at Hibbing School.

A violin! I'm sure it was a violin I heard, but the old man at the auditorium said the organ can reproduce every sound except a violin. My stomach balls up and starts doing somersaults. I, in turn, get up and start pacing the floor.

Joe rubs his temples, probably to purge my jumbled emotions from his head. When I stroll close by him, he reaches out and grabs my arm. "What are you not telling me, Anna?"

"Do you know anyone who plays the violin?" I anxiously bite my lip, not sure if I should look further into this.

"Why?"

I turn to Hannah and speak in English. "Du yu know some-von who plaz violeen?"

"Albert Einstein," she immediately answers, her face beaming at being invited into the conversation. "He learned to play at the age of six. Wherever he traveled in the world, he always took his beloved violin with him. He even named it; he called it Lina," she prattles on.

At the mention of that name, Joe's guilt skyrockets, making my gut muscles twist into knots. I think I'm going to puke. *Buckle up and hold on tight, Tatiana, you're rapidly rocketing toward self-combustion.*

"Anna, tell me what happened to you," Joe pleads in Russian. Kneeling down in front of me, he reaches out and holds my hands like he's about to propose marriage. "Please, let me help you."

Flashbacks of Viktor proposing jump at me. Unlike his offer, this one I want to say yes to ... if only I could. *Kaboom!* Another shock wave of raucous feelings jolts through me, rolling over with the force of a two hundred and forty-ton mining truck, figuratively flattening me to the ground. I stare upward in wide-eyed disbelief. What is happening?

"What are you doing, Joe?" Hannah protectively settles herself beside me and fixes an angry scowl at him. "You certainly can't be proposing. Are you threatening her? Anna looks like she's just seen a ghost."

Ghosts! The haunting pain from the Hibbing school auditorium stretches out its gnarled fingers to tap my shoulder. Terrified, I make a snap survival decision.

"When I was at the school, I heard music coming from the auditorium. It was a violin, but the old man there told me the pipe organ can reproduce every instrument except the violin. I followed him toward the stage, and about two-thirds of the way down the aisle, an agonizing pain seized me. I still feel it here." I point to my gut, my heart and my shoulder. Then I wave my hand all around, almost knocking Hannah in the head. "It keeps *haunting* me."

Joe rises to meet me eye to eye. Instead of discrediting my foolish claims, he recounts tales of the school's haunted past. "There are lots of rumors about the auditorium, a girl with disabilities and the first stage manager dying inside of it, someone slipping off the second-floor balcony and a student being killed by a falling chandelier."

My heart's pounding wildly and my pulse is following suit. "Is any of it true?" If Joe really did grow up there, he should know.

"Yes, a girl with disabilities did die there, and local legend claims that in the 1960s, the first stage manager died in seat J47, which is about two-thirds of the way down the aisle toward the stage. Some claim his ghost still haunts that seat, because while all the other seats in the auditorium fold upright, J47 always stays down, even when it's empty."

Poof! All the air leaves my lungs and I plop down on the floor. "Do *you* believe any of this?"

"I don't believe in ghosts," he bluntly says, seating himself across from me.

Hannah, refusing to leave my side, also sits down and wedges herself partially between Joe and me. "I'm going to stick by you tighter than a hair in a biscuit, Anna, and if need be, I'll be after you like a duck on a June bug, Joe. I don't know what's going on here since I can't understand when you two speak Russian, but I know I don't like it."

"Sic your duck on the biscuit, Hannah, you'll have better success that way." Joe motions for us to leave. I grab his hand and we pull each other upright, but I lead us over to my bed. I don't want to wait the few minutes it would take to get outside to resume this conversation.

"I don't believe in ghosts either, Joe. My life is influenced by the decisions of others around me, not by ghouls or spirits," I firmly state, struggling to quash the irrational fears circling within me. "What do you believe in then?"

"I believe our reality is created from what we directly sense. However, our perceptions are so limited that we only detect only a tiny fraction of the world around us. There are many other realities out there."

Just great, I can't even handle this one. "Are you suggesting I'm detecting things outside the normal range of human perception?"

"It's possible. Human vision only detects thirty-five ten-thousandths of a percent of the entire electromagnetic spectrum. Factor in our linear view of existence, where most people see infinity as starting at zero and going in one direction forever, and what we are actually aware of is minuscule. If we take a look at the flip side of reality, it's quite possibly that the system we live in is nonlinear, and infinity could be a series of repeating spirals. We might even exist in multiple states at once, only we're incapable of perceiving them."

This conversation is getting totally out of hand. I need to narrow it down from infinity ... but I kind of want to know where he's going with this. "Are you implying I'm feeling things from a different reality?"

Joe turns his head and powers down. I quietly slink around to see his face. His eyes are closed, and he's so completely immersed within his internal contemplation that he doesn't appear to notice me. He remains in this trance-like state for several minutes before opening his eyes and rejoining my reality.

"Where were you?" I ask.

A slight grin crosses his face. "While the tales of your Hibbing trip are not within the normal bell curve, if you think I haven't been in this room the entire time, you're not even on the chart, Anna."

"You're the one insisting there are other realities. Where were you?"

"In Pader, Uganda, in a bleak, one-story concrete building with eleven mattresses on the floor. There was a sixteen-year-old boy, only half the size he should be, tapping a cup on the concrete for water. It was his only lucid communication. Everything else was nonsense."

"Uganda?" Why in the world did he end up in Uganda? The sixteen-year-old boy isn't the only one communicating nonsense.

"The Nodding disease starts with nodding. Next, when the children eat, they feel cold and the seizures start. The kids stop growing, stop talking and ultimately the disease destroys them." Joe's tone drops a notch, as if he recognized after the fact that what he's telling me is not what I need to hear. If that was his conclusion, he's right.

"Are you trying to destroy *me*? Your way of helping sucks!" The opposite of Joe, my voice takes on a higher pitch, although not anywhere close to Hannah's. *Hannah!* Has she been studying us all this time? I prefer to be the one doing the observing, not the one being observed. Spinning around, I see her perched on the corner of her bed, staring right at me with an intense look on her face. What does that mean? Is she wondering how she can help me or is she thinking I'm beyond her capabilities and considering calling in additional backup? *Note to self: Don't get Hannah involved in anything more than cursory conversations.*

"Scientists from the Global Disease Detection Operations Center tested for dozens of potential causes, but couldn't find any," Joe rambles on. "The source of the disease still eludes the world's best epidemiologists. Many local residents believe the disease results from evil spirits of the dead. Rebels slaughtered many people in their compound around the time these children were born."

"Hold it right there. You just told me you don't believe in ghosts."

"What I'm trying to explain, Anna, is there are things in this world that are unexplainable."

"And you couldn't have said that to begin with instead of spouting off convoluted tales about children suffering from a disease with no known cause and no way of saving them? How in tarnation did you suppose that would help me? No, never mind, I don't want to know." *Another note to self: Don't ask Joe to explain things.*

"It's the association I made," he sheepishly admits. "But they're not all dying. Sodium valproate, an anticonvulsant medication, now provides some relief."

"I don't want to hear any more about that!"

"To the mind that is still, the whole universe surrenders," he instructs.

Unbelievable. "Are you seriously throwing wu wei at me? Like I have a snowball's chance in full sunshine of pulling that off, as you so clearly pointed out to me. You're a frying pan naming the kettle burnt, Joe." *Yolki-palki!* I've spent far too much time in the presence of Hannah and her disjointed sayings.

"At the center of your being, you have the answer; you know who you are, and you know what you want," Joe continues coaching me.

"Well, Lao Tzu obviously wasn't acquainted with my life," I yell. My stomach is now a twisting nest of Gordian knots, and I have no idea how to untangle them. Solution? Flee the scene. I hastily beat feet for the door.

"Why do you think we exist, Anna? Albert Einstein—"

I crush my hands against my ears, barrel into the hallway and down the stairs. I'm not going to let this bloody emotional connection take control of me. I'm not going to let myself fall prey to Joe's gravitational pull, become trapped in his terrifying tumble from orbit, plunge through the atmosphere, plummet toward Earth and finally burn up in a fireball! *Agh.* Curse Hannah's ludicrous repertoire of phrases that has invaded my brain, and curse this blasted bond with Joe.

Chapter 20

Joe
Just over three years ago

"Jesus, Lina, what happened to you?" As soon as I say it, I instantly know what happened to her, and a burning rage rips through me. That damn drunken bastard! "Did you call the cops?"

Her silence tells me no.

"I'm not going to let him do this to you, Lina." Her swollen lip and bruised cheek are deepening the open wound in my heart. As I turn to go confront him, she seizes my arm.

"No, Joe, no. He lost everything. He just needs more time."

"No, he doesn't." Furious, I pull away from her and shove my way through the crowded high school hallway.

"Joe, stop!" Lina races to catch up to me and tugs on my sweatshirt. "Please stop!"

The incredible fear in her voice cuts through me, and I stop.

"Promise me you will stay away from my father. Promise me, Joe."

Joe
Friday night, October 5

I rush down the stairs after Anna. Even though I'm still royally pissed about her going up to Hibbing and digging around in my past, I shouldn't have set her off half-cocked into the night. Racing out the door, I spot her running in the direction of the arboretum and sprint to catch up. It doesn't take long for me to overtake her. Patiently, I keep pace with her until the top of the hill overlooking Stewsie Island, where she half collapses and cries out like a wounded wolf. I cautiously take ahold of her hand. She's sweating, and her fingers are cold and clammy. Apparently, filling her head with tales of ghosts, children dying and evil spirits wasn't one of my better plans.

"Why isn't your senior photo in the yearbook?" she demands.

That's what she wants to talk about? After all the discussion of realities and perceptions, she only wants to hear about a picture in a lousy high school yearbook? "It's a long story, Anna."

"It's a story I need to hear," she insists. "Because I have to know if I can trust you."

"And a picture will do that?" I utter in total disbelief.

"Why can't you ever answer a simple question?" She exasperatedly pulls her hand away, sits down and sullenly mopes, pulling her knees up to her chin.

I reluctantly sit beside her. The night sky is partially overcast, and the harvest moon is casting an eerie light on the water, augmenting the queer mix of emotions churning between us. Anna shimmies over until our shoulders touch. She starts shivering, so I wrap my arm around her and pull her in closer. There's a pleasant lavender scent in her hair. She murmurs something I can't quite catch, then snuggles her head against me. "Have you ever felt like your entire world turned upside down, Joe? I did, after my mama died."

I think, in her own obscure way, she's trying to reach out to me, but I don't want to talk about mothers dying. I switch to science. "I read an article today about the sun's magnetic fields flipping. The polar fields exchanged places."

"Exactly!" She sits bolt upright. "That exact flip happened in my life. I started experiencing *other* people's emotions, and I don't mean secondhand, like when someone laughs or cries with a character in a movie. I mean I was literally living them." She waits for some sort of validation, or at least an acknowledgment to her grand revelation.

I don't want to screw this up again and have a redo of her wildly racing off, so I keep quiet.

Anna picks up where she left off. "I became trapped inside my father's nightmare of misery. To survive, I crawled deep inside myself, shutting off the outside world, but it didn't help. One day, it just ended. All the emotions vanished and I stopped feeling the sorrows, joys, anxieties and loves of life. I stopped being human." She forlornly stares up at the moon as her deep pain seeps into every corner of my being.

"From my side of the fence, Anna, your emotional generators are working just fine." More than fine enough to totally mess me up.

"Listen to me, Joe." She pounds my thigh. "I can't cope with the confusing, overwhelming mass of tangled emotions that the world generates, especially you. It's beyond my current skill level, and you need to do something about it!"

I need to do something about it? *I'm not the source of this well of despair you keep falling into, Anna. And do you have any conception of the torture you're inflicting upon me when you have these gigantic meltdowns?* She needs a bit of a reality check. "This isn't a one-way feed. I'm forced to experience all of your hysterical emotions too. Has it ever occurred to you that it might be a good thing you've avoided feelings all these years?"

Her eyes are blazing, almost glowing in the dark. "How can you possibly be angry with *me* right now?" she shouts. "I'm the one that should be furious at you. *You!* Because of you, I can't keep this latch in my mind closed that locks out emotions. Whenever you're near me, I have to battle a constant onslaught of feelings that I've spent most of my life without. It's totally excruciating and totally incapacitating!"

Shit! Shoot me now. I smack my head, trying to knock out the absurdity of our insane situation.

"Nothing to say about all that?" she growls, then hangs her head. "Why do people say lightning never strikes the same place twice? It's not true."

"You're right," I agree, glad we're off the topic of me being the sole cause of her apocalyptic problems. "The Empire State Building gets struck by lightning about a hundred times a year, and the Space Needle in Seattle is hit so often it has twenty-four lightning rods."

Anna perks up a smidgen. "Instead of lightning, people should say that meteorites never strike the same place twice."

"Meteorites don't need to strike the same place twice. They cause far more damage. Currently, there are more than fourteen hundred potentially dangerous asteroids and meteoroids orbiting near Earth, and any one of them is large enough to end humanity." While this is not the most uplifting of topics, it beats discussing the multiple ways I'm ruining her life.

She lets out a snarl of annoyance as a warning to stop, but I don't care anymore ... actually it's that I do care, far too much, and dealing with her anger is better than dealing with her other emotions. So, I continue full bore ahead. "The annihilation of our world could potentially arise from any one of those shining beacons of light you so fondly gaze upon, Anna."

Damn. Her mood plummets. Why couldn't she stay angry? How do I get her out of this funk? I'm sure as hell not kissing her. Maybe if I attack 51 Pegasi b, it'll get her mad enough to fight back? I'm already regretting what I'm going to say, but I can't think of another option. Heck, one more bad decision won't make much difference in my life, there's already a special place in hell reserved for me.

"Back in 1908, the Tunguska asteroid exploded over a remote region of Russia destroying hundreds of square miles of forest. Do you know where the radiant for that asteroid came from, Anna?" I don't wait for her to answer. "It came from the constellation Pegasus, home of 51 Pegasi b. That uninhabitable rock you so fondly worship is actually the origin of massive destruction."

Instead of infuriating her like I had planned, my attack on her treasured childhood talisman backfires, plummeting Anna even deeper into a brooding, desolate state. Will I ever figure out what makes her tick?

With nothing better coming to mind, I continue talking about meteors. "The explosion of the Chelyabinsk meteor produced the largest infrasound waves ever to be recorded. The Antarctic monitoring station detected them from ninety-three hundred miles away. The explosion was touted as the shock wave heard round the world. Have you heard of that phrase, Anna? It's from a poem written by Ralph Waldo Emerson. 'By the rude bridge that arched the flood, their flag to April's breeze unfurled. Here once the embattled farmers stood, and fired the shot heard round the world.' The poem commemorates the beginning of the American Revolutionary War. It was the catalyst for—"

"Shut up. Just shut up, Joe!"

Finally, she's fighting back.

Chapter 21

Many scientists believe we understand each other not by thinking, but by feeling. Experiments at the University of Birmingham provided evidence that a minority of the population, when they observe another person in pain, experiences not only the emotional component of the feeling, but the physical component as well.

Tatiana
Friday night, October 5

"Shut up. Just shut up, Joe!"

What? He's happy that I'm getting angry. "You are a totally dysfunctional, completely messed-up being!" I almost send fire flaming out of my mouth as I yell this.

"Goodness gracious, great balls of fire," Joe mutters, seemingly randomly yet seemingly purposefully.

Goodness gracious, great balls of fire! That's another of my grandpapa's all-time-favorite expressions. He uses it so much you might even say it's his trademark. Does Joe know my grandpapa?

Get a grip, Tatiana. I jiggle my head, but that only causes Hannah-talk to bleed into my frequencies. *You need to pay the piper before you meet your maker.* I rattle my brain a little more, trying to clear the reception, but this time Ms. Mizuikova's words jumble their way in.

Do you realize how stupid and irresponsible it is to think with emotions, Anna? They can get you killed. "Why did you say 'goodness gracious, great balls of fire,' Joe?"

"I could have said bada bing bada boom, but those words don't have quite the same impact. 'Goodness gracious, great balls of fire' seemed the quintessential response after our talk of meteors, during which I had to endure every one of your accompanying emotional explosions."

That's not the type of answer I wanted. "*Where* did you learn that expression?" I target my question more precisely.

"There are numerous references for it: Jerry Lee Lewis' song Great Balls of Fire, the centuries old Chinese use of flaming projectile weapons, and jumping back a millennium, the Holy Ghost taking the form of tongues of fire."

Why do I even bother? All this bickering is only making the doubts we already harbor about each other grow larger.

"Leave," I order through gritted teeth. He doesn't budge a millimeter. It seems I'll need to employ additional tactics to get rid of him. "*You* are the focal point, the spring board and the epicenter of my emotional issues, and I'm so sick of you purposely irritating me or redirecting conversations if they involve something you want to avoid. You need to leave, now."

"And what is it you think I'm trying to avoid, Anna?"

How I'd like to know the answer to that. *Just get out of here, Joe. I want you to go* ... yet I want him to stay. "Let's start all over. What I was originally trying to explain is that the major rift in my life is our relationship. Whenever we're together, my emotions go haywire and I can't function. If I don't figure out why this is happening and how to stop it, I won't be able to remain at Carleton." And if I can't keep a clear head, I may end up dead.

"I am not the sole cause of your problems, Anna."

Why can't he understand what I'm telling him? "I can't figure this out on my own, Joe, and it's getting worse every day. Recently, even when we're separated, I still feel the bond between us. It's getting stronger and stronger, and it's going to take both of us to stop it." I explicitly spell it out for him.

Clenching his jaw firmly shut, probably to keep himself from saying something else that will piss me off, he starts swearing under his breath. After a particularly long string of foul words, he bites his lip so hard I think he's going to draw blood. "I'm drawn to you too, Anna, and I don't want to be. I'm so damn tired of continually fighting this attraction to you."

At long last, a glimmer of understanding. That's why he's been purposely goading and angering me. To push us apart.

"No matter how hard I try to stay away from you, I can't, because I'm haunted by this feeling that you're in danger."

He feels it too. How much weirder can this get? In answer, torrents of droplets, filled with the superfluous baggage we recklessly added to our tenuous situation tonight, rain down over me. Pulling up my hood, I take cover. I'm trapped in the middle of a storm where there's no easy way out; neither of us has any grand escape plans. At this point, I'd settle for a mediocre plan, but we don't have one of those either.

"Who is Eleanor Engel?" I ask, simply needing to think about something else for a while. That was a bad move. The heart-wrenching guilt that so frequently torments Joe rips into me with an especially cruel intensity.

"It's a long story," he wearily replies.

"I think it's an intricate part of whatever is forcing us together, so it's one you need to tell me."

"It's a story I don't want to share, Anna," he flatly refuses.

Why are our lives plagued by so many secrets? And why are there so many strange coincidences surrounding us? Using Hannah's technique, I start adding them up on my fingers. Number one—Joe's use of my grandpapa's preferred sayings. Number two—his ability to speak fluent Russian. Number three—his knowledge of my favorite chocolates, Alyonkas. Number four—his vast knowledge of space, including Kepler. Number five—his memorization of Kepler quotes.

"What are you doing, Anna?"

Number six—his awareness of my school report on Pyramiden. Granted that's a bit of a stretch, but why would he have such in-depth information about that obscure town? I've recounted too many connections between us for them to be pure chance. Resetting my hands, I put both my fists in front of my face.

If you're going to smack me, Anna, you don't need to count the reasons why. Go ahead and give it your best shot. I won't try to stop you."

Why does he know all this information about me? I've already determined that it's impossible he's a sleeper agent specifically planted here for me, because the timeline is way off. But I still think he could be a sleeper agent who's here for another reason and stumbled upon me. Yet, if that were the case, why does he keep trying to push me away? Sometimes, it almost feels like he's trying to warn me that *he* is the danger. "Is Eleanor Engel dead?" I ask.

Bada bing bada boom! Joe's raging demons savagely attack me, sending both of us dangerously close to nuclear meltdown. I've got to do something immediately. "Joe, listen to me. Over and over again, your presence has forced me to relive the death of my mama. Oddly, I'm beginning to accept that the grief will always be there." For a brief instance, Joe and I are wholly of one mind, living the exact same experience. *Keep talking, Tatiana, it's working.* "I'll never stop grieving for her, because I'll never stop loving her. The two are interwoven and will always be part of me." *Ofiyet'!* I'm actually helping *him* cope with emotions.

"If I hadn't met her, I'd probably be dead or in prison now," Joe confesses.

"Met who?" I anxiously cross my fingers, hiding them under my legs, and hoping he'll tell me. I assume it's Eleanor Engel, but assuming things is a bad way to operate.

"The summer after our junior year of high school, her mom was killed in a car crash. Her dad survived, but soon afterward, he started drinking, lost his job and started hitting her. The bastard hit Lina!"

Lina! The name pulses through me with a shocking energy.

"The fucking asshole promised her he wouldn't do it again, and she made me swear I wouldn't go after him." Joe's voice is starting to falter. "He hit her again, and I couldn't uselessly stand by and do nothing. Lina knew that, and before I had a chance to confront the son of a bitch, she did the only thing she could think of to stop me. She disappeared." This last bit gets caught in his throat. Joe's devastating self-condemnation is so vivid and sharp it feels as if it just happened seconds ago. "It's been three years, thirteen days, four hours, seven minutes and twelve seconds since then. She left me on September twenty-second, 2009."

September twenty-second, 2009. Lina left Joe the same day that Viktor was killed! Another coincidence? My astonishment is quickly smothered by another wave of Joe's guilt. Unable to endure his suffering any longer, I tenderly turn his head toward mine and touch our lips.

The kiss deepens, sparking a blinding heat to burn between us. Grasping the back of my neck, he strokes my skin with his fingertips, leaving a scorching trail of desire with every tingling caress. Our flaming passion incinerates everything, and we gladly give in to the pure unhindered bliss we're experiencing. I'm happily swirling in this cocoon of joy when, out of nowhere, Joe suddenly breaks away.

His emotions are all over the place and his hold on his self-control is almost nonexistent. "Why did you do that, Anna?" His voice is harsh.

"I couldn't sit by, with your form of hell streaming through me, and do nothing," I answer honestly, phrasing it in his own terms so he won't misconstrue my intentions.

"Is that the only reason?" His face is flushed with passion ... and confusion ... and yet more guilt.

"You can feel what I'm feeling, Joe."

"No, I want to hear you say it." He slams his fist on the ground.

"Yes, it's the only reason. To soothe your heartache is to ease my own. We're bound together, whether we like it or not."

Chapter 22

Incas believed they were born into a world where every living being experiences the energy of every other living being as if it were their own. This concept, Kawsay Pacha, sees the world as alive and responsive, viewing the cosmos as a vibrating field of pure energy frequencies.

The human brain is also a field of energy frequencies, containing roughly one hundred billion neurons. Each of these neurons constructs up to ten thousand pathways with other neurons, making the number of brain-activity permutations greater than the number of particles in the universe. All this brain activity generates a measurable electromagnetic field, which in theory has the possibility of carrying information to other unconnected neurons.

This all leads to very unconventional questions. Is it possible to sense each other's mind? Is it possible to touch each other's consciousness? Are we all connected?

Joe
Friday night, October 5 to Sunday, October 7

As I head back to my dorm, after walking Anna to hers, my lips are burning and my gut is sick with guilt. One part of me is mourning for Lina, the other is gripped by a fierce desire for Anna. It's ripping me in two. I've got to find a way to deal with this … with her. Ignoring Anna didn't work, trying to push her away didn't work and toughing it out isn't working either. What's causing this damn connection between us, and how can I put an end to it?

Once in my room, I immediately start searching the internet for information on sensory input. In no time, I'm inundated with data. Snakes use infrared to interpret their world, bats navigate with air compression waves, ticks (which are blind and deaf) utilize temperature and butyric acid, and dogs (with over two hundred million scent receptors) employ smell. Since different animals engage differing methods to observe their fragments of reality, maybe certain people can too. As I keep reading, something really strange blows me away. Sensory substitution, substituting one brain signal for another.

"You writing a report on sensory substitution?"

I involuntarily recoil, startled that Lance was able to sneak up behind me and I didn't notice. My sensory perception sure sucks at the moment. "No, I'm not."

"Shame, it's an intriguing subject. Anyway, you should still check out BrainPort. It's a small device that sits on the tongue and changes visual signals into electrotactile signals. Over time, BrainPort allows blind people to "see" their surroundings by using their tongue. With his eyes closed and his tongue out, Lance wanders around until he bangs his shins against his bed frame. "Guess I need more practice, but isn't that a wild concept?"

"Yeah, wild," I mumble.

"It does makes *sense,* Joe. Since all the signals going to the brain are electrochemical, it doesn't matter where they come from, as long as the brain knows what to do with them." Approaching my desk, he reaches into his pocket, pulls out some coins and throws them at me. "Here, have some extra *cents*. Don't worry, your brain will figure out what to do with them." He laughs at his corny joke.

"Funny, Lance. Do you know anything about ESP?" I might as well tap into his knowledge since he seems to be well read on the topic.

He puts his fingertips on the sides of his temples and spins them in circles. Then he hums and stares straight ahead like he's in a stupor. "Ah, the famous sixth sense. I see a building with a sign and bright red letters, Top Secret Stargate Project. I'm sensing more. It was part of the US government's psychic arms race with the Soviet Union. Uh oh, it's starting to fade. Wait, there's a huge pile of money. Millions of dollars for ... extrasensory perception and—"

"Snap out of it, Lance. I don't need a séance. You can return to your normal knightly duties."

He gives a mock bow, then begins backing away, but he can't quite fully disengage. "I would be remiss in my duties if I didn't mention the movie *The Sixth Sense*. It's about connecting with dead people. There's this quirky kid who—"

"Dismissed, Lance." I do, however, look up the Stargate project. It's a real thing. The study lasted for over two decades and one portion of it focused on psychoenergetics, a mental process where one person perceives and influences another person who's separate from them in space or time. *Whoa.* That really jives with the wacky connection between Anna and me. I read on, hoping it will lead to a way out of this weird connection between us. Nope, nothing conclusive. After thousands of trials, Stargate ended with researchers claiming the results show humans have the capacity for ESP and the Defense Department disclaiming it.

I didn't ever expect to be doing this, but I start examining articles about empaths, individuals who claim they take on the emotions of others around them. Often times, it's difficult for empaths to discern other people's emotions from their own and they can easily become overwhelmed. The more I read, the more I'm forced to consider that maybe there's some validity to this paranormal stuff, because Anna sure seems to have it. When you add in Lance's comments about our brain's electrochemical signals and the fact that our thoughts leave behind electrical signatures, psychoenergetics isn't so far-fetched. But I'm not ready to buy into it yet.

The rest of the weekend, I spend every moment of free time learning more about psychic phenomenon, hours and hours scrutinizing research papers, reports and data. Now it's Sunday night and I still have no clue how to manage what's going on between Anna and me.

Rap, rap, rap. She's at the door. I *feel* her, but I don't want to see her, so I ignore the knocking. For five straight minutes, she continues pounding, then it stops.

"Hey, Romeo, you have a visitor," Lance announces after unlocking the door and coming in.

I don't want to interact with Anna, I don't want to be bombarded by her discordant emotions and I sure as hell don't want to be tempted by my attraction to her. She senses this as she tentatively takes a few steps toward me.

"Aren't you going to introduce us?" Lance asks as he reaches into the plastic bag that's wrapped around his wrist, pulls out an Almond Joy and throws it in my direction. It ricochets off my computer screen and lands on the keyboard. Extracting another chocolate, he politely offers it to Anna.

Accepting his bribe, I do introductions. "Lance, this is Anna. Anna, this is Lance. You're both obsessed with the Kepler space mission, so go somewhere else to share your infinite wisdom."

Accustomed to my terse remarks, Lance glowers at me for a short moment before turning his attention to Anna. "Did you hear the sad news? NASA admitted the Kepler space telescope is beyond repair. Two of its four reaction wheels are broken."

Anna doesn't comment. Her focus is on something else. Me.

Lance is confused by our lack of interaction and glances back and forth between us. But Lance being Lance, he only falters slightly before jumping back into discussing his current pet project. "The Kepler team is continuing to analyze the data already collected and are expecting to find thousands of new exoplanets. Maybe they'll even find the long-awaited one in a habitable zone of a sun-like star, the Goldilocks World!"

"Hate to break it you, Lance, but *Goldilocks* is a fairy tale." I want to shut him up or at least get him to stop talking about the Kepler Mission. Once he gets geared up on these citizen-science projects, he rambles on forever. Living the CitSci life is Lance's endeavor, not mine.

"We will find Goldilocks!" he exclaims, loud enough to make both Anna and me flinch. "Even if Kepler doesn't lead us to the Goldilocks world, there are other options such as the LSST, the Large Synoptic Survey Telescope in Chile, which will be taking a ten-year time-lapse photo of the universe, and TESS, the Transiting Exoplanet Survey Satellite, which will scan the entire sky for exoplanets, and the James Webb Space Telescope, the most powerful telescope ever created which will use infrared vision to study every phase in the history of our universe, including the formation of solar systems capable of supporting life on planets like Earth."

Longest run-on sentence ever. "What the *Keck*, Lance?"

Anna shares a brief smirk with me. Even though she probably didn't follow all of what Lance was rapidly spooling out, she recognized the names of the telescopes and therefore got my quip. My remark was lost on Lance though.

"What the heck is wrong with you now, Joe?" Walking over, he grabs back the Almond Joy that's still lying on my keyboard.

Darn, I should have eaten that right away. "I was only pointing out that you completely overlooked the Keck Observatory in Hawaii," I politely clarify, endeavoring to get the candy back. My justification appeases him enough to toss the Almond Joy at me. This time, I tear it open right away and pop it in my mouth.

Having a guaranteed block of uninterrupted time while I'm busy chewing and swallowing, Lance theatrically sweeps his arms out, signaling for us to huddle together. Anna takes a step back and I remain seated. Lance takes it in stride and resumes his pep talk. "NASA's asking for our input. Even though Kepler can't continue its original mission, its light-gathering mirror and electronics are still working. They're requesting proposals for future Kepler science missions. We should all work collaboratively and—"

"No." I put a quick halt to this.

"But think what our three brilliant minds could achieve together," he argues.

"No, I'm not doing any more extracurricular projects, Lance. That's your thing. I'll draw it out for you." I sketch a face with a tongue sticking out, above a sideways-written K, enclosed within a little pentagon with the letters CCCP written on top.

He studies it for a moment. "What's that supposed to mean?"

Anna skirts over to look. "Znak kachestva," she says.

"Correct." I tap the symbol. "Znak kachestva is a Russian mark of quality created to represent outstanding workmanship, but it became synonymous with poor workmanship and low expectations. That's precisely what you'd get with me working on one of your projects. So why don't you—"

"Knock it off," Anna angrily snaps at me in Russian. "Lance isn't the source of our problems, so quit antagonizing him." Seizing my shirt, she tugs us so close together that our noses almost touch. This intimacy isn't lost on Lance who moves into my limited field of vision, purses his lips together like he's kissing, then mouths the word "girlfriend" at me.

"I need your help picking up something that requires identification," Anna urgently whispers in my ear.

"And why do you need me for that?" I ask, mildly curious.

She hesitates and peeks over at Lance. "Can we talk somewhere else?"

"No." I break out of her grasp. "I don't want to be alone with you, Anna, and besides, Lance doesn't understand Russian."

"I need you to make an ID card with Ms. Mizuikova's name on it," she absurdly blurts. "It's vitally important and I can't do it myself. Ms. Mizuikova is watching me like a hawk, and it's getting almost impossible to slip away from her." Anna's oppressively somber, brown eyes stare at me expectantly.

What the hell is she thinking now? "Medusa's a snake, not a hawk; she's a low-lying reptile, not a high-flying predator." *And I'm not getting suckered into more of your issues, Anna.*

Instead of arguing, like I hoped she would, she quickly plants her lips on mine, playing dirty and knowing the intimate contact will intensify our attraction to each other. I instantly break the kiss, but not in time. The full depth of her despair and fear work their way into me, causing the now almost primal need to protect Anna to exponentially increase.

Lance raises his eyebrows and gives a sly grin. "You two are an item."

Enough of this shit. Getting up, I pull on my coat and head for the door.

"Would you help me if my name was Lina?" Anna yells from across the room.

Her question stops me dead in my tracks. *Damn it, damn her, and damn this connection!* I want so badly to walk away from Anna and never look back, but I can't. She didn't yell that out of spite, she did it out of desperation, from a well of loneliness and a lifetime of isolation. Reluctantly, I turn to face her. "Why do you need this identification?"

"To retrieve some letters," she lamely answers.

"And why do you need Medusa's identification for that? Give me all the gory details." I can only imagine where this is leading.

"My father set up some ridiculous security modus operandi, and *my* private mail is being forwarded to a PO box in Wisconsin registered in *her* name."

This is getting more convoluted by the minute. "Then just ask her to pick up your mail."

"I can't." Her voice is getting panicky.

The feeling that Anna's in danger is becoming overpowering, undermining my resolve. I sense myself starting to cave. "Why can't Medusa get them for you?" I irritably ask.

"It's a long story."

Very clever, Anna, throwing my own words at me. I copy her game plan and repeat her earlier words. "It's a story I need to hear."

"It's a story I don't want to share," she cunningly continues the playback.

"Then it's a story you're going to have to finish without me." I hurry out the door, grateful that for the most part my roommate is content to be an NPC in my play calling, unlike Hannah who insists on being a participating character.

"Joe, wait. I'm sorry I asked you to do that. Pozhaluysta, hear me out," she implores as she chases after me.

Saying please isn't going to work this time, Anna. Spinning around, I confront her with the question I've had since we first met, and she has yet to answer, "What do you really want from me?"

"I … I …" she stammers.

I don't have a choice, do I? If I turn her away and something terrible happens, I'll be reliving this decision forever. I'll never forget.

I resign myself to helping her, but with my stipulations. "There's an interesting story about the Space and Naval Warfare Systems Marine Mammal Program, Anna. A dolphin that was trained to hunt for mines discovered a one-hundred-and-thirty-year-old Howell torpedo. It was one of the first self-propelled torpedoes used by the US Navy."

"Does this mean you'll help me?" she optimistically interrupts.

"Only fifty Howells were ever built. The only other one known to still exist is in the Naval Undersea Museum in Keyport, Washington."

"Is that a yes?" she presses for clarification.

"It signifies that you never know when long, hidden dangers will come to light under the most peculiar circumstances. I'll help, Anna, but you need to agree to my terms and do this my way."

Anna's eyes widen with some sort of revelation. "Viktor," she whispers before she can stop herself.

Viktor? Who the hell is Viktor?

Chapter 23

Tatiana, age five

It's late, and the house is cold. I silently creep over to the shadows by the doorway. Grandpapa is still here.

"Goodness gracious, great balls of fire!" he exclaims. "That damn diamond mine has been nothing but trouble since Boris and I first invested in it. For seven months of the year, the temperatures are so cold that the workers have to use jet engines to burn through the permafrost, and the entire mine has to be covered to prevent the oil from freezing and the steel machinery from shattering. During the brief summer, it's nothing but mud."

"I wouldn't consider six hundred million dollars in annual profits trouble," Mama argues.

"It is when you're being prosecuted for tax fraud." Grandpapa angrily whips back. "I should have known from my days in the KGB to stay away from that wretched Mir diamond mine. The Adámas will be a thorn in my side until the day I die."

"The Adámas?" Mama questions. There's total quiet. "Who's the Adámas?" she repeats.

"Forget you ever heard that name," Grandpapa warns.

Papa steps in. "Who is the Adámas? What danger does he pose for us? Will he pull our family inside the Kremlin's crosshairs that are currently focused on you?"

After a long pause, Grandpapa quietly says, "He was an engineer who stole a fortune in Silver Bear diamonds from the Mir mine. I was involved in the investigation. The Adámas was never found."

Adámas is an ancient Greek word from which the word diamond is derived. It's means unalterable or unbreakable.

Tatiana
Two years ago

"Those are grandpapa's problems, not mine, Father. I've got nothing to do with that stupid diamond mine. And no one listens to what I say anyway."

"They are listening, Tatiana, and your grandfather's problems are our problems too. You need protection and putting an ocean between us will not end this."

"Yes, it will, Father."

"No matter how disconnected and isolated you think you are or will be, you're still linked to this family. We are all connected."

Tatiana
Sunday evening, October 7

"Viktor," I whisper before I can stop myself. *The Adámas and the silver bear diamonds,* the little gremlin on my shoulder spits in my ear. Where did that thought come from? That name hasn't been spoken since my grandpapa moved to England.

Long hidden dangers will come to light under the most peculiar circumstances. Joe's statement hits the mark dead center. One thousand yards, standing position, iron sights, and he nails the target.

What are the odds?

The odds are not in my favor, not with the dead/alive Viktor bearing down on me. I have to get to that post office and intercept any more letters before Ms. Mizuikova gets them or I'll be ensnared within my father's security web and shipped off to London, which was where *he* wanted me to attend university.

For a third time, I try to get a definitive yes out of Joe. "So, you'll help me?"

"Not until you tell me exactly why Medusa can't pick up your mail."

My two inner voices start arguing. One says, *Yes, tell him* and prods me to let Joe be the lone exception. But the other, the cold, cutting voice that has kept me alive all these years is declaring, *No, you can't trust him.*

"Why, Anna?" Joe harshly asks.

Perhaps if I disclose disconnected, misleading bits of backstory, just enough to gain his cooperation, I'll appease both my inner voices. I rush ahead and offer him a doctored accounting before I change my mind. "The letter I received was from a student I knew in secondary school. After I was assaulted, my father wanted me as far away as possible."

"That doesn't explain much of anything about why Medusa shouldn't retrieve your mail."

I begrudgingly confide a little more. "Her main purpose here is my safety, but I will not have her snooping around in my private affairs. If she tells my father that I'm getting letters from this guy, he'll insist I relocate again."

Joe rolls his eyes at me. "If her real purpose here is your safety, then Medusa's already told your father about the letter."

Blin. Why does he have to be so perceptive? "You're right, she did tell my father. However, the letter is signed with the initials KA, not his real initials. That's why it hasn't set off DEFCON 1 procedures."

I feed him more twisted details of the story. I only hope I can keep all this straight later.

"Why is it signed KA and not with his real initials?" Joe isn't as easily put off as my father was, and that's saying a lot, because it took quite a bit of effort to fend him off.

"KA stands for King Arthur," I truthfully explain. "It was a nickname he used, claiming that like King Arthur, he would always return if needed." Ironically, this was a promise made by both Viktor and King Arthur shortly before their deaths.

Joe gives me a scornful look. "What else are you not telling me?"

This blasted connection! I can't hide my anxiety, and Joe's becoming too adept at using this link to interpret my reactions.

"Let me read the letter," he uncompromisingly orders.

Now I've talked myself into a corner. The letter wasn't signed with the initials KA, which actually is Viktor's pseudonym. *Wait a minute!* Why didn't I think of that before? Viktor would definitely have written down KA. No more need to speculate about the possibility Viktor might still be alive and tracking me down, leading who knows else to my hiding place. Yet, I can't let Joe look at the letter. As soon as he sees it, he'll know I've been deceiving him. "I can't show it to you."

"That's quite a tale you're feeding me, Anna," he scathingly states, then strides off.

"I'll let you see the letter tomorrow morning on the ride over to the Eau Claire post office," I yell after him. Desperate times call for desperate measures. Without Joe's help, I won't be able to intercept any more correspondence in time to prevent deportation by my father.

Joe doesn't slow down at all. If anything, he's moving faster.

"I won't be imprisoned by my father's ridiculous security measures again. I'll take off and disappear if I have to." Joe's unexplainable need to protect me takes over and his shoulders slump in surrender. I immediately press him while he's still susceptible. "The identification card needs to be done by tomorrow morning."

"I can't go with you tomorrow, Anna. I'm not cutting class for this."

"You don't have to come with me." In fact, it's highly preferable that you don't accompany me.

"I'm not letting you go by yourself. This Viktor guy could be waiting there for you."

I don't want Joe tagging along and interfering. Well, better to ensure I have the ID card first, then find a way to ditch him. "The post office opens at eight and it's only a two-hour ride, so we can easily be back here in time for your class."

He eyes me critically, his misgivings won't quit. "I'm not agreeing with your methods, Anna. I'm only doing this because I know if I don't, you'll go blindly racing off by yourself without any thought of the dangers around you and ... and we can't get caught doing this," he ridiculously tacks on.

"Do you really think I need to be told that?"

He pulls out his cell phone, snaps a photo of me and gives instructions to be ready by five thirty.

"I'll be outside my dorm by five twenty," I reply, attempting to keep the giddiness out of my voice while my inner self performs a happy dance. A happy dance? The last time I did one of those was when my mama told me she would take me to the planetarium.

But, am I happy because Joe's going to help me or because we'll be spending many hours alone together?

Chapter 24

If there is space between objects, we consider them as separate and distinct. But experiments performed during recent decades have shown that something we do over *here* is subtly entwined with something that happens over *there.*

Tatiana
Early Monday morning, October 8

Dark, damp, dismal and abysmal. That about sums up our situation as we venture out at five thirty in the morning for a possibly worthless, possibly perilous and certainly fractious endeavor.

We wearily walk in silence to the parking lot while a tangled mass of emotions closely encircles us. Not wanting to drive, I hand my key fob to Joe, get in on the passenger side and input the address into the navigation system. Joe puts the car in gear, and we both brace ourselves for the two-hour drive to Eau Claire.

I look through the playlist on my iPod, select a song and adjust the volume on the car's sound system. *Sittin' here, with nothing to do, I think of you. Confusion, confusion, it's nothing new.* The lyrics dovetail with my current state of awkward affairs. Seconds into the song, Joe slams on the brakes. "No music," he firmly states.

"Are you serious? What are we going to do for the next two hours?"

I always listen to music in the car.

"Turn it off or I'm out of here," he says in total seriousness.

Your face is all aglow, I want to say hello.

Click! Joe unbuckles his seat belt, opens the door and starts getting out.

"All right, already, I'll turn it off." Huffing my aggravation very noisily, I put away my iPod. How am I going to make it through this drive with no music? I didn't bring any earbuds.

Three minutes! Three long minutes with nothing to distract me, three painstaking long minutes of being in the close confines of the car centimeters from Joe is all it takes for me to realize what a huge miscalculation I've made. The depth and breadth of this bond between us has been growing, and it's to the point where it functions even when we aren't in close proximity. Being trapped right next to him has pushed the intensity meter into the red zone. What was I thinking when I agreed to this?

Watch out, Tatiana! An intense emotional downdraft sweeps over me, then around me and every which way like ping-pong balls in a lotto drawing. Amid this chaos, disorganized thoughts of the many perils around me flash in and out. A vision of Viktor's letter spontaneously explodes in my hands, setting off countless microbursts in my brain like swarms of fireflies blinking on a windless summer night. I, too, blink rapidly, trying to extinguish the multitude of little blazes. Closing my eyes tightly, I stretch my taut muscles and try focusing on a rock in a river.

Humph, wu wei. That's useless. From out of nowhere, a vision of Aunt Elena riding a Nimbus 2000 broomstick appears. She's hysterically waving at me when she suddenly screams and plummets downward. Reaching forward to catch her, I end up banging my head on the dashboard. *T'fu.* Well, at least the bump has cleared my scattered thoughts.

I guardedly peer around. We haven't even gotten out of Northfield yet! My pulse is already racing, my nerves are on edge, my blood pressure's rising and I'm feeling nauseated. Taking a sideways glance at Joe, I notice he's having just as hard a time coping as I am. His muscles are locked up, his entire body is rigid and he has such a death grip on the steering wheel I'm worried he's going to snap it in two. At this rate, we'll never make it out of Minnesota.

Another image of Viktor pops up. He's in his uniform, bloodied and dying. A woman is bent over him, weeping and cradling his head in her lap. There's a name written on the back of her shirt, *Lina.* Viktor died on the very day Lina left Joe!

You've got to distract yourself, Tatiana. "Why are there penguins everywhere, Joe? Doesn't that bother you Carleton Knights?"

He has no idea how to respond to my absurd question, and more torturing minutes drag by. "You're a Carleton Knight now," he finally responds, probably in self-preservation. "And the penguins aren't everywhere, only in the library, on some T-shirts, on a mural in—"

"The penguins bother me," I abruptly shout as we pass by a cemetery. American cemeteries are so mundane, no colorful fences, no decorative benches to sit on and no portraits on the gravestones. The straight, organized lines of tombstones add to their boring appearance which is quite unlike the maze of haphazardly placed graves in a typical Russian cemetery.

"I'm lost," Joe mumbles in an almost inaudible tone.

He's obviously not talking about penguins anymore, and I doubt he's referring to a Russian cemetery. "Lost? How can you possibly be lost? Aren't you following the navigation system? We're on a major road with markers every so many kilometers, there's no excuse for getting off course."

"Not that type of lost, Anna. Lost as in pushed off track by wildly changing emotions, disorientated by a crazy conundrum of futile research about telepathy and now thrown into this goat rope of a trip."

There are all sorts of hidden innuendos in there, and I'm not setting foot anywhere near them. "Why are there penguins everywhere, Joe? You haven't answered me." I have no idea why I keep asking this, or maybe I do. I don't want any more silent time with *wildly changing emotions*.

Neither does Joe, because he restarts pontificating on penguins. "Laurence Gould, Carleton's fourth president, was the chief scientist and second-in-command for Admiral Richard E. Byrd's first expedition to Antarctica. After a grueling two-and-a-half-month, fifteen-hundred-mile trek on a dogsled, he brought back Oscar the emperor penguin. Oscar is on display in the Gould Library. That's why penguins are part of Carleton history, Anna."

Yolki-palki! What are we doing? Sitting here discussing penguins when I have danger all around me. I revert to the survival training that's been drummed into me and start scanning my surroundings. "Joe, stop!"

He slows the car and a large black cat fleets across the street. "Another Joan Feynman lives to see another aurora," he comments with a grim smirk.

"You remembered." He remembered I name all my cats after Joan Feynman. A warm tingling fills me.

"How could I not have remembered?" he irritably mutters.

I'm confused. There are lots of reasons why he wouldn't have remembered ... unless he's been instructed to memorize everything he can about me. It's too quiet in here again. What topic should we talk about next? "Have you noticed the abnormally high number of coincidences that surround us, Joe?"

"Synchronicity—people, places, patterns or events that reoccur in time. The word 'synchronicity' refers to the wheels of time and—"

"I don't want you to be a walking encyclopedia. I want you to think about what I'm asking you."

"Okay, Anna. Synchronicity. You meet someone who touches your soul, and that person comes into your life over and over again. You begin to feel a joint destiny and start thinking with your heart instead of your head. Is that better?" He looks over.

Eureka! I feel like a little girl who's so giddy she can't contain herself and jumps up and down with excitement. I feel like a debutante at her first ball in that first mystical moment of that first dance, swirling in delight. Joe actually skipped over all the extraneous, distracting sidebars he normally throws at me and got right to the heart of the matter.

"Think about it," I say. "We both have a vast knowledge of space, we both speak fluent Russian, we're both socially isolated from the world around us, we both have hidden secrets that haunt us and now we're both hardwired into each other's emotional network. Do you think we've been forced into each other's life for a reason?" *Or were you forced into my life for a reason?*

He grips the steering wheel even harder. "The word 'synchronicity' was coined by Swiss psychologist Carl Jung in the 1920s. He was captivated by the idea that life was not a series of random events, but an expression of a deeper order. He believed synchronicity shifted a person's egocentric thinking toward a greater wholeness."

So much for skipping the distracting sidebars. "I don't care what Carl Jung believed. I want to know what you think."

"Do you truly want to know what I think, Anna?" He shifts uncomfortably in his seat and keeps his eyes fixed on the road.

"Yes, I just told you that," I impatiently state.

"I think my view of reality is evolving more rapidly than I can adjust to."

He's starting to grasp that elusive wisp of a notion, that nebulous conception about the interconnectedness of … of what? The answer remains just out of reach. "How is your reality evolving?" I ask.

"After Carl Jung talked with Einstein and Wolfgang Pauli, he became drawn to the parallels between synchronicity and quantum mechanics. Both of these point toward the interaction of matter and consciousness, a link between the objective and the subjective world."

I'm frustrated that he's slipped right back into scientific-lecture mode, but I attempt to follow what he's saying because I have the odd feeling he's guiding me toward something significant.

"Currently, there are two competing theories, locality and nonlocality. In simple terms, locality states that objects separated by large distances *can't* directly affect each other. Nonlocality, also known as quantum entanglement, states that objects, even ones separated by billions of light-years, *can* directly affect each other." He pauses to make sure I'm still paying attention. "With quantum entanglement, there's an instantaneous connection, The rate of speed is limitless. Yet Einstein proved that the speed of light is the maximum speed. The two theories are in direct conflict."

"I don't want a lengthy physics lesson, Joe. Speed it up and get to whatever point you're trying to make."

He hits the accelerator, and I'm jammed into the seat. "Gravity, good suggestion, Anna, but it doesn't help me make my point." He slows down to just above the speed limit. "During Einstein's time, there was no way to evaluate quantum entanglement. Then in 1964, John Bell's experiments recorded the spin measurements on pairs of entangled electrons. The results supported Einstein's predictions."

"Hurrah," I sarcastically cheer. "Now skip to the very end of this lesson, the very last sentence on the very last page. What does it say?"

"Let me explain it a different way, Anna. I'll put Einstein's theory in terms of you and me. When something happens to one of us, it changes the universe in such a way that the other is immediately affected. The space between us doesn't matter, the connection still exists. But, and it's a big *butt,*" he gives me a wink, "for quantum entanglement to work, we need to have a common history. That's what binds our futures together."

What is our common history, Joe? How can I get that out of you? "Why are we entangled?"

"Do you mean in a manner similar to *spooky* action at a distance? That's how Einstein referred to it."

I give a slight nod of my head. After my recent visit to the Hibbing School, I'm not so quick to dismiss spookiness.

"Consider the human body, Anna. It develops from one cell, ergo, all the cells in our body are connected. If we consider that one cell, which originated from atoms forged billions of years ago in the cores of far-flung stars and—"

"No more scientific dissertations." I've reached my far-flung limits. Gazing out the car window into the darkness, David Bowie's song *Space Oddity* flings into my thoughts. *Hmm.* The stars do appear different.

"The sky does seem different, Anna."

It's eerie when Joe seems to knows what I'm thinking. "Why did you say that?"

"My telepathy only activates when you're on audio. You were singing, not very loudly, but enough so I could hear you."

Why must I do that when I'm around him? Note to self—Keep my mouth clamped closed when I'm pondering anything in the presence of Joe. Another note to self—Don't ask him about anything to do with physics. I have a feeling the do's and don'ts list labeled, Joe, is going to be quite long by the time this trip is over.

"Colonel Hadfield sang that song when he was on board the International Space Station. He thinks sort of like you do, Anna. Hadfield claimed that from space you see the whole world as 'us,' and a sense of connectiveness fills you. Although, 'connectiveness' is not technically a word. He should have used 'connectedness.' "

"Only you would comment on that, Joe."

"No, Lance would have too." He chuckles. "Back to Hadfield. There are numerous other words he could have used, affinity, kinship, similitude, consanguinity, coalescence, connection, interdependence, amalgamation, conflation. Now there's an interesting word for you, conflation. It's when the identities of two individuals, concepts or places that share some characteristics appear to be a single identity. Have you ever played the game The Minister's Cat, Anna?"

Where in bloody tarnation does he come up with this obscure stuff? His ramblings are way more off-the-wall than my penguin question.

"No? Let's play a guessing game instead. Are you ready?"

That's got to be better than listening to him drone on and on about vocabulary words or physics lessons. "Okay, Joe, go ahead."

"First one. 'A human being is part of a whole, called by us the universe.' Identify that quotation."

"Albert Einstein." I take an educated guess, knowing Joe's strong affinity for him.

"Spot-on, Anna! Next one. 'Maybe it's all men an' all women ... The human sperit, the whole shebang. Maybe all men got one big soul ever'body's a part of.' "

I have no idea.

"Time's up. That was from *The Grapes of Wrath* by John Steinbeck. Moving on. 'Ninety-five percent of the mass of the universe is something we can't even see, and yet, it moves us. It draws us. Studying science, you realize the *connectedness* of all things.' " He nods to himself approvingly. "This individual uses the correct word. Connectedness, not connectiveness."

Never mind that, Joe, let me answer. "Natalie Batalha," I proudly state.

"First prize for the young lady in the front seat. Let's move further back in time. 'Friendship is a single soul dwelling in two bodies.' "

"Aristotle," I state. I'm actually starting to enjoy this game.

"Very impressive. Let's bump it up a notch. 'Not only higher forms of life but also the smallest insects survive based on a recognition of their interconnectedness.' "

Hmm. This one is more difficult.

"That was His Holiness the fourteenth Dalai Lama. Try again. 'Enlightenment for a wave in the ocean is the moment the wave realizes that it is water.' "

Another one I've never heard before. I'm starting to lose interest, and the reference to water which leads me to think of wu wei adds a bit more distaste.

"That was Thich Nhat Hanh, a Vietnamese Buddhist monk. I'll make this next one a little easier for you. 'The cosmos is within us. We are—' "

"That's enough!" A spell of silence is initiated. I quickly realize what a bad thing that is and restart a conversation. "How do you remember all that crap, Joe?"

"It's my curse," he bitterly says.

"I distinctly remember you telling me that you don't believe in ghosts, so I doubt you believe in curses. On that opinion, I'm with you. Our lives are controlled by our actions and the actions of others around us, not by some sort of voodoo," I state this firmly, even though I'm not as sure of myself on matters of the supernatural as I once was.

" 'We all dance to a mysterious tune, intoned in the distance by an invisible piper,' " Joe recites with a touch of reverence.

Where is his mind off to now? Well, at least he's talking about music. "Who said that?" I inquire.

"Albert Einstein. Do you know that the brain generates a measurable electromagnetic field which in theory—"

"No, I don't, and I don't want to hear anything about it." I glance at the dashboard clock. It's only six thirty! You've got to be kidding me. We still have an entire hour left to drive. I give us only fifty/fifty odds we make it all the way to Eau Claire.

Chapter 25

Joe, age twelve

"It's time you learned how to fight," Kluge tells me.

"I already know how to fight," I spit at him.

"Your black eye and busted lip tell me otherwise."

"You don't know anything, I flattened both of them. I can take care of myself."

"That's not good enough, Joseph." *I will need you to take care of Ellie also.* "You're smart, strong, agile and predatory in nature. You should be able to outfight anyone you meet."

"I can outfight anyone I meet, and I don't need a weak old man like you giving me advice on fighting."

With unexpected speed, Kluge pins me against the wall and twists my arm up behind my back. "Never underestimate your enemies, Joseph."

Joe
Monday morning, October 8

It's only six thirty! I'm going to go completely batshit before we reach Eau Claire. How did I let Anna talk me into this? There must be something we can discuss that's not going to zap her emotions into hyperdrive ... Got it.

"Did you go to a dacha growing up, Anna?" Kluge once mentioned his dacha to me. It was one of the few reminiscences he shared with me that brought a brief smile to his lips, and it was one of the rare occasions when he wasn't mean and ornery, which was almost all the time.

"Yes, Joe." She heaves an exasperated sigh.

There's no way she can be frustrated with this subject already. All she's said about it is the word yes. "Describe it to me."

"I don't remember much. We never went there again after my mama died."

"Tell me what you remember." Any conversation is better than none, and from Kluge's telling of it, a dacha is similar to a summer cabin for a Minnesotan. It's a refuge where you can get away and let life slow down for a bit. So, it might bring up some happy memories for her.

Anna starts curling her hair around her finger. That's a mannerism I haven't seen her do before. She pensively nips her lower lip, then her face relaxes. "The outside of the cottage was painted a faded blue color and the trim on the windows was white. At the edge of our front yard, there was an old, weather-beaten picket fence."

Yea! She's having some cheery thoughts. "Keep going," I encourage.

Her face scrunches up in deep thought. "Outside the fence was a long wooden bench, and in back of the dacha was an old outhouse that my grandpapa built. I used it once when I had to pee so bad I couldn't wait. It was an extremely traumatic experience due to the large population of spiders in there. I was terrified some were living inside the hole and would crawl all over me when I sat down." Even though her nose and eyes crinkle up at the memory, her mood's still improving. "We had fruit trees. Mama would make jams and pies from them. Raspberry was my favorite."

Raspberries don't grow on trees, Anna. Chances are slim she'd appreciate my humor, so I keep it to myself.

"Sometimes I'd go down the road with my mama to help pick raspberries. They were so warm from the sun that they almost melted in my mouth."

"What else did you eat?" I ask, wanting to keep this upbeat tempo.

"Melons. I especially love watermelons. After dinner, we'd all sit out back around a large wooden table, and my father would cut up the melon. Even though I was the smallest, he'd hand me one of the biggest pieces."

This is great. I'm getting to know the *real* Anna. All this talk of food, though, is making me extremely hungry.

"Do you have anything edible in this car, Anna?" If not, we're stopping. I don't care how much she'll protest about the delay. She reaches into the back seat, grabs a plastic grocery bag and pulls out a package of Corn Nuts. Not my favorite thing, but it'll do in a pinch. "Anything else in that bag?"

"Four more packages of Corn Nuts." She rips open the one she's holding and starts crunching away. "I make sure to have a ready supply of these on hand wherever I might be. Munching on them in front of Ms. Mizuikova makes my day. Occasionally, it annoys her so much she just ups and leaves." Anna offers me a handful and I enthusiastically join in, relishing the thought of Medusa writhing in abhorrence of the sound.

"What did you do for fun at the dacha?" I ask between mouthfuls.

"Swim and run." She ponders this question for a bit longer. "Sometimes Mama would take me mushroom hunting in the woods. We'd both be so thrilled when I spotted some."

"You got excited about mushroom hunting?" I can't see that.

"Yes, Joe, I did." She impudently juts out her chin. Suddenly, I can see it and can't help chuckling over the image of a tiny Anna ripping mushrooms out of the ground.

"Glad you find it so amusing," she snaps, but she's not really upset and continues reliving the pleasant memories. "I looked forward to everything about the dacha, even loading up the car until it looked like it would explode. One summer it was packed so full we barely had room for the cats."

Tempting, so tempting. *No sarcastic cat comments, Joe.* "How long would you stay there?"

"All summer, or at least my mama, my brothers and I did." Anna gives a wistful sigh. "Time passed slower at the dacha. It almost felt like another world, especially at night when I'd stare up at the stars." Pushing a button, she opens the sunroof, or at the moment the star roof, and gazes upward.

Drawn in by her comments, I do some reminiscing of my own. "Someone once told me that the dacha is the soul of Russia, the only place where one can feel free."

Her head instantly snaps back down and she intently stares at me, unable to hide her deep interest. "Who told you that?"

Someone who made me swear on my mother's grave to never mention his name.

"What? Now is when you decide to stop talking? Ever since we got in this car, you've bombarded me with in-depth physics lessons and quotes about consciousness, became a walking dictionary, then a game show host and just recently started hounding me with questions about my dacha. But when I ask you a single thing, you shut up?"

"I can't tell you, Anna. I made a promise." And I won't break it. "Although, since you mentioned hounding, let's talk about dogs. On second thought, cats would be a better subject for you." Discussing cats should *purrr*k her up. "After hearing your comment about the cats dying at Pyramiden, I immersed myself in the world of cat lovers." I can't believe I subjected myself to that torture, but Anna's influence makes me do things I'd normally never consider. "Guess what I found out?"

"Lots of physics lessons about cats," she sarcastically says.

"Actually, yes, even an entire book filled with them called *Falling Felines and Fundamental Physics.* It was inspired by the fact that if you drop a cat off a building, it will land on its feet. This natural feline instinct has provided valuable information in the fields of mathematics, neuroscience and human space exploration. I carried out a few such feline experiments in my youth." *Uh-oh, wrong thing to say, Joe.*

Anna's figurative claws start extending, and I quickly drop this subset of cat talk. "That's not what I intended to tell you. I discovered Nyan Cat, an 8-bit, computer-generated, Pop-Tart-shaped kitty that got over seventy-nine million views!" And there are millions of these moronic cat videos on YouTube. This truly baffles me, for after watching dozens of them, I couldn't find any redeeming value. "Kitty videos are a colossal waste of time." *Shoot.* I got caught up in my thoughts again and wasn't filtering. I sure as heck didn't mean to say that to her.

Anna's whole face darkens, so much for reliving the happy memories. "Who gave *you* the right to decide for the rest of the world what's a colossal waste of time?" Her back arches and her imaginary tail grows twice the size it should be. "Coming from a prodigy who wastes most of his waking hours on football, your opinion holds no merit and should be thrown away with the kitty litter. I've got a news flash for you, Joe, cats have won the internet." Her hackles are way up. Good thing she not an actual cat, or I'd have scratch marks over most of my body by now.

Yet, maybe that comment wasn't such a bad thing. Her spunky attitude is an improvement over the brooding, mournful mood she was in when we started this drive, and a good heated debate might be exactly what we both need to survive this trip.

"I discovered something else beyond belief, Anna. The Walker Art Center hosted an Internet Cat Video Festival and over ten thousand feline fanatics attended." Why a reputable museum would do this is an enigma to me. "It's unconscionable that countless hours of creativity and productivity were wasted this way." *Crap!* I did it again, broadcasting what I'm thinking. Anna's malady of doing this, thinking out loud, seems to be contagious. At least I'm aware of this bug in my programming, unlike Anna, who doesn't notice when she's inadvertently voicing her thoughts.

She's getting really pissed off with me. I should back off, but I can't resist because I want to find out something. Was she or wasn't she? "I wonder what type of person would stand for hours at a cat-rat festival to have their photo taken with Dusty the Klepto Kitty preposterously dressed in a police uniform?"

"It was a prison uniform," she hisses in a menacing voice.

Oh, Anna, you're such an easy target. I know what he was dressed in. "You were there, you attended this farce."

"It's a far cry better than watching grown men in tights chasing after a small pigskin." Razor-sharp weapons emerge from where her fingernails should be. If I weren't driving, I'd be protecting my face right now. I guess I took this a little too far, like the time when my chance encounter with a stray cat, followed by my brilliant idea of hiding him in my third grade classroom, left me with deep scratches up both arms. Back then, there were no warning bells going off in my head, signaling me to abort. I hear them now, loud and clear, yet words keep pouring out of my mouth. "I'm simply explaining to you that cat people have an irrational view of the world."

"That's what you've been trying to do all this time? Prove that I'm irrational? Congratulations," she screams. "You win."

I don't feel like I've won anything. In fact, I definitely feel like I lost.

Chapter 26

I must seek in the stars that which was denied me on earth.
Albert Einstein

Tatiana
Monday, October, 8

Why does Joe do this? What purpose does it serve him to infuriate me? He keeps glancing at the clock every few seconds. *That won't make the time pass faster, Joe.* After an especially long sideways look, he holds up his right hand with his index and middle finger extended in the peace sign. " 'I'm free, yet I'm trapped. For those observing from the outside, this life seems so normal.' "

He's quoting Kuzma! The Siberian cat sensation on the internet.

"Kuzma seems to have a higher level of intelligent thinking than most of the other internet cats," he says in a repentant voice.

Is this an apology? If not, then what is he up to now?

" 'I'm cursed and haunted by my past. Voices beckon, yet I comprehend nothing. All their communications are meaningless to me. I yowl and howl for help. No one understands.' " Joe reaches over and places a hand on my knee. The hairs on the back of my neck stand up. Even though Kuzma's a cat, his predicament strangely parallels my own. Another coincidence?

" 'High on the stern, Joseph his stand, held out his hand. This message bear: Bring holy peace and beg Anna's relief.' "

"Are you offering a truce?" I ask.

"Connectedness," he simply states.

"I don't get the connection."

"Like Roman relationships, ours seems to fluctuate between waging war and offering cease-fires," Joe informs me. "So I borrowed some words from Virgil, an ancient Roman poet."

"Oh, I thought you were apologizing," I flatly state. "While that's an interesting analogy, Joe, I see our relationship more as shifting between war and passion."

Turning his head ninety degrees, Joe eyes me for so long that I'm about to reach for the steering wheel. "I'm sorry, Anna," he says, then looks back at the road.

Oh, my tumultuous, blue eyes of Siberia. One moment, you're pressing the knife to my throat, the next you're begging forgiveness.

Joe swerves over, parks the car and turns off the engine. We've arrived at the Eau Claire post office. "Let me see the letter," he brusquely demands, completely altering course from his seconds-ago apology.

I don't have any choice. Unless I let him read it, he won't help me. Reaching into my jacket pocket, I take out the letter and hand it over.

Ever since you left, I've felt so alone.

Without you, the world is such a dismal place.

The days are excruciatingly long and the nights even longer.

I can't wait to see you again.

Noticeably missing are the T at the beginning of the note and the V at the end. I cut them off last night, because as hard as I tried, I couldn't figure out how to explain them to Joe. He'd be expecting an A for Anna and a KA for King Arthur. Even though it's obvious I cut the paper, and there's no doubt he noticed, Joe doesn't ask me about it. Instead, he methodically refolds the letter, hands it back to me and stoically sits there, trying not to show any emotion. Still, I feel what I'm sure he wants to keep hidden. Jealousy.

"It's 8:02, the post office should be open," I state, hoping to get him moving.

Joe takes out his wallet and removes a surprisingly legitimate-looking replica of a student ID with my picture and Ms. Mizuikova's name. As long as it's not inspected too closely, this might work. After handing over the ID, he firmly instructs me not to speak in English. "In fact, don't speak at all," he tacks on.

I manage to stifle my retort, not wanting to discourage his cooperation with me on this important errand. We enter the post office and ... wait. Only 8:04 in the morning, and there's already a long string of people winding around the room. Russia isn't the only country where one idly bides time in long government lines. At last, it's our turn and we approach the counter, side by side.

"Ze haz forgot her keys andt vantz gedt mail from PO box," Joe says in a hideous attempt at a Russian accent.

The postal worker stares at us, then asks me, "What's your PO box number?"

As I'm about to reply, Joe forcefully kicks me in the shin to shut me up. I'm rapidly taking a dislike to his version of the plan. "What's your PO box?" Joe asks me in Russian.

"One-five-one-four," I angrily state.

"One-five-one-four," Joe repeats in English.

"I need to see some identification." The postal worker is still staring at me.

Playing my role, I impatiently wait until Joe asks me in Russian for my ID. I remove it from my purse and place it on the counter. While the worker looks it over, Joe begins volunteering unsolicited information. "Ze vaz franteek ven lost keys. Came du me en paneek.

I no zink eez end of vorld. Only meaz ze vill nod drool over ledder from boyfren."

All right, that's enough. Putting my full force into it, I kick Joe's shin. The resulting grimace on his face and the harsh clenching of his teeth provide me some form of satisfaction. The postal worker returns my ID and says he'll be right back. The second he's gone, I confront Joe. "What happened to your don't-talk philosophy? Could jealousy have anything to do with your lapse in judgment?"

"I was keeping his mind occupied while he was checking your identification," he smugly answers.

"Bullshit!" That's not at all what you were doing. You were speaking those words at me, not him.

The worker emerges from the back and makes a point of handing the two manila envelopes to me, not Joe. *Hooray!* Triumphantly, I march out the door and happily strut down the sidewalk toward the car. *Whoosh!* Joe snatches the envelopes out of my grasp and shoves them deep in his jacket pocket with the crass authority of a dictator. How could I have been so careless? I let myself get foolishly elated with the adrenaline rush, and now look what happened.

"Give Me My Letters," I vehemently threaten. Clamping onto his arm, I put my full weight into trying to stop him, but he just pulls me along the sidewalk, like a mule dragging a cart. I think my shoes are making skid marks on the concrete.

"Don't make a scene," he chillingly instructs.

Before I can toss some choice words at him, a heavyset man steps in front of us, blocking our way. He's wearing a fedora hat with a wide brim that's pulled down to conceal his face, a long gray coat and black leather gloves with gray stitching. "That's no way to treat a lady," he booms in a baritone voice.

"Mind your own damn business," Joe spits back.

This isn't good, Joe's not one to back down. We need to leave here immediately before this gets out of hand and attracts police attention.

"Apologize to her," the man orders.

Joe levels a lethal stare at the guy and tells him to get the hell out of his way. I tug on Joe's sleeve, hoping to avert the escalating conflict. "We *need* to go," I urgently prod him.

Untangling himself from me, Joe shoves the guy out of the way and resumes walking toward the car. *Whew!* I breathe a sigh of relief. That was a close call. *Oh, no.* The man grabs Joe from behind. "I told you to apologize to the lady," he firmly repeats.

This time, Joe isn't able to rein in his temper and slams the fellow into the wall of the post office building, getting right up in his face. "You haven't got a clue how I treat women. Stay the fuck away from us." The situation is rapidly escalating. The man takes a swing at Joe, but Joe reacts faster, sidestepping and giving a punch of his own, a brutal left uppercut to the jaw. He doesn't stop there; Joe throws three more hard punches. A right to the gut, a left to the man's exposed ribs and a right to the face, breaking the guy's nose. This isn't just a street brawl. Joe's been trained to fight.

"Stop it!" I put the full force of my emotions into the words.

Startled, Joe jumps back a step. Had anyone else yelled at him, I don't think he would have heard, let alone reacted. His focus was solely on his adversary. Luckily, my near-lethal dose of ratcheting alarm shooting directly into him, seizes his attention.

Score one for this weird connection.

Chapter 27

Joe, age five

I'm sitting in a corner on a small stool in the principal's office. She's already finished yelling at me, and now we're waiting for my mom to arrive.

The principal gets up to greet her. It makes me feel good that my mom's taller than her. "Sorry to have to bother you again, Mrs. Carleton, but this is the third time already this month that Joseph has pushed another boy down. I have to suspend him." My mother nods, then motions for me to join her. "I've tried reasoning with him, but he is obstinate and uncooperative," the principal continues.

"She didn't listen to me, Mom. All she did was yell."

"Let's go, Joe," she says in a tone that means don't you dare argue. Once we're in the car driving home, my mom's face softens. "What happened?" she asks.

"That bully Harry took Peter's snack away from him again, and Peter didn't do anything. He just let him take it. It made me so mad, because it's not right Harry can do that, and no one stops him. So I took it back, then I smacked him." I look over at my mom. She's staring straight ahead at the road. I search my brain for a word to call Harry that will make her understand. "He's ... he's intransigent, Mom, and the teacher wouldn't listen to me, neither would the principal."

Tears are starting to fall down my face.

My mom pulls the car over to the side of the road, pulls a tissue out of her purse and dries my tears. "Oh, my little Einstein. The world's just not ready for you yet."

Joe, age sixteen

I walk up the steps to Kluge's house and find the door ajar. Something's not right. He never leaves the door open. Cautiously, I enter and spy Kluge's body lying in the hallway. He has a large gash on his head and blood is still pulsing onto the floor, pooling around his body. Before I can check to see if he's alive, someone grabs me from behind and pins me to the wall.

"Where's the money?" the attacker demands in Russian.

"What money?"

"Don't play stupid with me, or you'll end up like this foolish old man." He jerks his head in the direction of Kluge.

I push against him with all my strength and manage to grab my switchblade. Kluge gave it to me a few years back and insisted I carry it with me at all times, and I do, except when I'm at school. "Never underestimate your enemies," I snarl as I stab the intruder as hard as I can.

He staggers momentarily, then comes at me with a knife of his own. We struggle, and I continue thrusting my blade into him. I think I'm gaining the upper hand.

"Drop the weapons!"

The words don't register with me, and I continue stabbing my attacker. Seconds later, I'm again slammed against the wall, only this time it's by a police officer. The intruder falls to the ground and loses consciousness. Later that day, he dies.

CHAPTER 27

Joe

Monday morning, October 8

Anna lays into me the moment we're back in the car. "*What* about not making a scene did you not understand? What is wrong with you?"

How many countless times have I heard that question? Tabulating all the instances people have lectured me about "what's wrong with me" could keep me occupied for the entire ride back. My dad alone could fill up pages and pages. "Do you want the complete list, Anna?"

"No, but I want to know where you learned to fight like that?"

"Leave it!" I can't mention Kluge's name to anyone.

"Don't tell me to leave it. Did it cross your mind for even a brief second the consequences your reckless actions might cause?"

"Did you consider for even a brief second why that man approached us at all and what type of threat he might be to you?" How can she not be mindful of this? It's incredible she's survived so long under the Russian regime while blindly ignoring the risks around her. I start the car and turn onto the street with Anna glaring daggers at me. A deathly quiet, twenty-five-minute ride ensues. Finally, we cross the border into Minnesota, and I exit at the rest area. Both of us are still barely containing our fury as I park the car.

"Give me my letters!" Anna rages as we come to a stop.

"After I read them," I shout back. If I'm going to get dragged into the middle of this mess and have my life on the line for her, I want to know what kind of trouble she's in.

"Give them to me!" She starts trying to claw her way into my pocket, and I manage to fend her off long enough to pull out the letters. Holding them as far away from her as possible, I inspect the first envelope. It's addressed to Anna Alkaeva, c/o Ms. Miesha Mizuikova, PO Box 1514, Eau Claire, Wisconsin. The return address is 20 Wildberry St, Proctor, VT. *Well, I got that part right, the previous letter was mailed from Vermont.* Ripping it open, I extract another envelope. This one is addressed to Tatiana, c/o Elena Proctor, 20 Wildberry St., Proctor, VT, with no return address. It's postmarked from Moscow. Inside is a sheet of nondescript white paper.

My darling T,

Your face is on my mind when I drift off to sleep.

You remain with me throughout my dreams.

You are my only refuge.

V

I open the other envelope within an envelope.

My darling T,

How I long for your touch,

To brush my lips against yours and revel in the sensation.

This separation is more than I can bear.

Promise you will wait for me.

V

"Not very long-winded, Anna, is he? So, who is V? I want the truth this time. Who wrote these letters?" I'm furious with her for deceiving me.

"I told you the truth," she insists.

"I want the whole truth, *Tatiana*." She reacts just as I expected, with a quick flicker of fear. Tatiana is her real name. "If this Viktor attacked you once before, and now he's tracking you down, your father wouldn't be ignoring this."

"Don't call me Tatiana." There's a touch of panic in her voice.

"Tell me who wrote these letters, Tatiana." There's no one around to hear me, so I might as well use her name to my advantage.

"Viktor ... He was a boyfriend of sorts," she reluctantly admits.

"Damn it, Anna. The *whole* truth," I firmly press. "I wasn't born yesterday."

"Funny, that's what I want from you too, the whole truth. Only I don't have any leverage on you ... yet," she clips back.

Her incredible rage is wrecking mayhem in my head. Controlling my own emotions right now is hard enough, having hers flung in makes it nearly impossible. Throwing the envelopes and letters at her, I get out of the car, slam the door and head to the men's room. I turn the faucet on full bore, blasting cold water over my bruised knuckles. It eases the throbbing in my hands, but doesn't do a damn thing for my temper. Why does my life always get so screwed up? Tossing some of the cold water on my face, I attempt to calm down. No chance of that. I give up and return to the car, determined to confront Anna again and get the truth out of her about "Viktor."

"Stop jerking me around," I shout at her the instant I'm seated. "We're not leaving here until you tell me exactly what kind of trouble you're in."

She stubbornly digs in her heels and retreats inside herself. The sky is darkening and a light rain starts hitting the car, cocooning us inside. "I learned to survive alone," Anna utters in a voice that's cold and distant. "Before my mama died, I prayed for her suffering to go away. I promised I'd be good, I'd be nice to my brothers and do whatever I was told. Foolish, foolish childish fantasies," she mocks her former self.

Her words burn deep into me. Her childhood wasn't so different from my own. " 'To see one's mother go through the agony of death and be unable to help—there is no consolation. We all have to bear such heavy burdens, for they are unalterably linked to life.' Albert Einstein had similar feelings as us, Anna."

"Why am I like this, Joe? Even though I live in the midst of people, why am I always alone?" she despondently asks.

I'm certain she doesn't want to hear more of my esoteric quotes, but I'm afraid of what will happen between us if I try to physically comfort her. And I can't rely on my personal wisdom to answer her, so I again let Einstein guide me. " 'The woman who follows the crowd will usually go no further than the crowd. The woman who walks alone is likely to find herself in places no one has ever been before.' "

"What if I don't want to walk alone anymore?" The depth of Anna's despair disintegrates my resolve to avoid intimate contact. Reaching across, I gently massage her tense shoulder muscles. She tilts her face toward mine, and the alluring emotions we've been fighting this entire trip take over. Tucking some stray hairs behind her ear, I lean in and kiss her sweet lips. Our blood pulses wildly out of control and our breathing becomes rushed. Unzipping her jacket, I slide my hand under her shirt and bra to caress her soft breasts. I want to touch her all over, and I want to love her all over, in every meaning of the word.

Da, da, da, da, da. A quick drumming on the car roof brings my barely functioning brain back to where we are. Looking out the window, I spot two young teenagers darting off. *A car parked in the middle of a rest stop. What are you thinking, Joe? Getting sexually involved with Anna? What in the hell are you thinking?* I start up the car and zoom us back onto the highway. Once cruise control is set, I glance at Anna. Her jacket is zipped all the way up to her chin, her face is firmly set at a ninety-degree angle to lock her vision straight ahead and she's valiantly trying to remain composed.

"Do you feel alone now, Anna?" I ask, not able to hide the resentment in my voice, most of which is frustration at myself. Why did I do that? Why did I kiss her again? Lina's still out there somewhere, and she's much more alone than Anna is. I need to keep my damn hands off of her and keep my damn mouth shut. *Don't say or do anything else, Joe. Don't risk it.*

It's a long ride back to Carleton with nothing to distract me. Soon, intense memories of Lina intrude, then things go from bad to worse and striking images of Anna juxtapose themselves against Lina. The clashing discord between the two is brutal.

"What are you thinking about?" Anna irately asks.

"You." *And Lina.*

"Think of something else."

That's not possible, not when I'm trapped in this vehicle inches away from you. I hunker down as another memory of Lina appears. This time, her face is cut and bruised from her father's beating. If only I had done something differently. If only I had done something sooner. If only I could find her.

Whether out of annoyance or for self-protection, Anna intercedes. "Viktor's dead. He died three years ago."

Unlike me, she doesn't have feelings of guilt or sorrow associated with her loss. In fact, she doesn't seem to have any feelings about it at all. *What the heck?* I don't know which one of us is more dysfunctional. Anna, who's almost devoid of passionate memories about Viktor, or me, who has a limitless supply of them about Lina.

"He died on September twenty-second," she announces.

The same day Lina left me! Her revelation stuns me for a moment. That is quite a coincidence. However, I still don't buy into fate or superior powers controlling us. There are far too many vicious, random, unconnected acts of terror in this world for me to believe in any overarching view of connectedness. "If Viktor didn't write those letters to you, then who did?"

"Don't know," she aloofly mumbles.

"You need to tell your father about this. You're in a lot of danger." And you definitely need someone else besides the viper looking out for you.

"I'm not telling my father," she resolutely snaps. "I've had threats on my life before. I know how to protect myself."

Why does she insist on doing everything single-handedly? "Let me help you, Anna."

"I can't."

"Remember what you said to me. 'You always have a choice.' I'll make your decision even easier by narrowing down the choices. Either tell me what's going on, or I'm ratting you out to Medusa; I'll fill her in on today's field trip and the letters we picked up."

Alarm flits across her features, then her expression hardens. "Tell me who taught you to speak Russian and where you learned to fight like that," she swiftly counters. "Or I'm ratting you out to Medusa and telling her everything I know, which she in turn will pass on to my father."

Damn. I can't have that, Anna knows quite a bit about me now. Point, counterpoint, stalemate reached and conversation stopped. Once more, with nothing to hold my attention, I'm sucked into my internal battles and bombarded by more memories of Lina and Anna. This time, Anna's face is foremost, staring at me with stone-cold features that are empty of any emotion, except for her eyes, where an unimaginable desolation bleeds through.

"Stop it!" she cries out. "Whatever thing about me you keep zoning in on, over and over and over again, knock it off."

Using every ounce of willpower I have left, I try shifting my thoughts. Words from Lao Tzu appear. *Watch your thoughts; they become words. Watch your words; they become actions. Watch your actions; they become habit. Watch your habits; they become character. Watch your character; it becomes your destiny.*

Destiny!

Chapter 28

No one anywhere can answer for you those questions and feelings that deep within them have a life of their own ... Do not now seek all the answers, which cannot be given because you would not be able to live them—and the point is to live everything.

Rainer Maria Rilke

Tatiana
Monday, October 8

Will I ever be free to be me, just me? Will my soul ever be able to sing openly?

Oh, no. Joe's torturing himself again, and me along with him. I don't want to speak with him anymore, but I can't suffer through any more of this either. There must be some safe subject I can bring up. *Aha.* There's the article I read on the 'astronauts' of inner space. That shouldn't be too contentious. "Some extreme divers recently explored underwater caves called blue holes which are—"

"I know what blue holes are, Anna, and that's not what we should be discussing. You're in grave danger," Joe admonishes me. "Let's figure this out together. When did you first meet Viktor? Maybe one of his friends is writing these letters."

I'm not going to talk about Viktor or those letters with him. "The lower layer of blue holes gradually loses all light and oxygen, perfectly preserving what's within it and forming a sort of looking glass into what life was like hundreds of thousands of years ago. On—"

"We need to focus on what's happening right now," Joe perfunctorily interrupts.

"No, *we* don't! I will handle my matters by myself." I set him straight then pick up where I left off. "On the cosmic calendar, hundreds of thousands of years are only seconds ago." Delicate bubbles of connectiveness encasing glimmers of understanding appear in midair, just out of my reach, and float there like hot air balloons ready to launch. I stretch out, longing to grab ahold of one before it drifts away.

"It's not all one big shebang, Anna. Shall I tell you how many seconds there are in those hundreds of thousands of years?" And just like that, the balloons are far up in the sky, their wisps of wisdom scattering to the winds. Joe clicks on the blinker and turns into a McDonald's.

"What are you doing?"

"What does it look like I'm doing? I'm starving." He pulls up to the drive-through and orders two Egg McMuffins, a sausage burrito, hash browns and a coffee. "Do you want anything?"

Yes, I want to understand this strange feeling I keep having about connectedness.

Joe waves his hand in front of my eyes. "Hello? Anybody in there? Do you want anything to eat?"

"Scrambled eggs, toast and orange juice," I reply. He repeats my order, drives up to the next window and hands over some cash. In short order, the food server shoves a bag of food and two drinks at him, which he promptly hands over to me. I sort through it, give Joe his sandwiches and place the drinks in the cup holders. With amazing speed, I wolf down my food, partially because I'm hungry and partially because I'm anxious to resume the discussion about connectedness, even though it's been a mostly one-sided conversation.

"At the bottom of a blue hole is a dark and desolate place without any oxygen or sunlight; life shouldn't be possible there, but it is." Suddenly, it's tugging at me with a particularly strong sense of urgency. "Life does exist there, Joe, and we're connected to it. At times, it seems like that connection is beckoning to me and I feel caught between two worlds, the one I know and another I'm just becoming aware of ... I need to protect it."

"Protect what?" he garbles through his partially open mouth before swallowing a large bite of burrito.

Protect what? The question rattles around inside me and the kernel of understanding slips away, again. The hot air balloons floated off without me. Deflated, I fall back inside the small space of my blue Prius.

"You're in danger, Anna. We both feel *that,* so stop diverting me with irrelevant matters."

"You do not have the final say in what is irrelevant. And why is it all right for you to divert me with extraneous reams of gibberish, but I'm not allowed the same privilege?" Even though I've never expected fair play in the game of life, I want it now.

"When I tell you something, Anna, it's not irrelevant, but that in itself is irrelevant. We have to focus on who's trying to hunt you down."

"No, *we* don't!" How many times do I have to hammer this into his head?

"Fine. Then don't expect any more help from me with your all-important errands."

Neither one of us speaks for the remainder of the trip. It's not very far, yet it feels like forever. When we reach the campus parking lot,

Joe swerves into an empty spot, tosses me the key fob and storms off.

I'm left sitting ... alone.

Chapter 29

Our prime obligation to ourselves is to make the unknown known. We are on a journey to keep an appointment with whatever we are.
Gene Roddenberry

Joe
Monday, October 8

I tear off as soon as we're parked, wanting to get as far away from Anna as I can. But the farther I go, the more my anger is overshadowed by doubt. What if Anna gets killed by this Viktor fellow and I could have stopped it? *Damn it!*

Reversing course, I run and catch up with her, then trail her up the stairs and into her room. Hannah's seated at her desk, making origami cranes, but her attention immediately zooms in on us. Anna's radiating hostility—Hannah's eyes narrow in concern. I try to hide my bruised knuckles—Hannah scowls at me. Anna opens her locked drawer and throws in the new letters—Hannah frowns. She's studying us like we're lab rats. Her expression ultimately settles on stern disapproval.

Obvious questions have formed in her mind. However, she's smart enough to discern that neither one of us will answer them, so she starts in by streaming from her ample supply of idioms. "Well, well, look what the cat dragged in. It appears you two were fighting like cats and dogs. I guess you're no longer the cat's meow, Joe."

Her perverse form of feline torture sucks. Remarkably, Anna's not in the least phased by Hannah's prodding. She's completely blocking her out. Why can't she do that with me?

"Cat got your tongues? You both look as comfortable as cats on a hot tin roof." Hannah pitter-patters about the room on her tiptoes as if the floor is steaming hot. "In fact, I'd go as far as to say you're jumpier than a long-tailed cat in a room full of rocking chairs." Not getting a noticeable rise out of either of us, she rummages deeper into her arsenal. "Okey-doke, there's more than one way to skin a cat." Using an imaginary knife, she begins "skinning" me, all the while screeching like a stuck pig.

Screw it! I can take any more of her crap. "Curiosity killed the cat, satisfaction brought it back. Beware, Hannah, I won't bring you back."

"Rah-rah, Joe. You're finally willing to play cat and mouse. But first, a correction, Curiosity was framed, ignorance killed the cat." She chuckles at her clever comeback. "So, where have you two cats been prowling around?"

"I don't want to be killed by ignorance, Hannah, do educate me. If I throw your cat out a car window, does it become kitty litter?"

A hurt frown crosses her face. She picks up a plastic bag which she then viciously shakes in front of my face. "Let the cat out of the bag, Joe. Why are you such a jackass?" Unexpectedly, she switches gears, sweetly patting my arm and crooning at me. "Cats are like music; it's foolish to explain their worth to those who don't appreciate them. You have my deepest sympathy, Joe." Turning tail on me, she focuses on Anna, who's busy shoving things into her backpack. "Your pet snake, Medusa, stopped by here extra early today. She was fit to be tied and got all catawampus when she found out you weren't around."

Hannah breaks out in a small grin. "That nickname, Medusa, tickles me pink."

"Bevar, ven cat andt mouse agree, grozer eez ruint," Anna feeds the bizarre form of *cat*egorically, *cat*astrophic, *cat*aclysmic communication, then *cat*apults herself out the door, leaving me stranded with Hannah.

"What's a grozer?" she asks me, disturbed by Anna's stranger-than-usual behavior.

"A grocer," I distractedly reply. I have to keep a close eye on Anna because she's not safe here, but I can't spend all my time trailing after her.

"What were you two doing this morning? She's more preoccupied than a cat covering crap on a marble floor."

That's because you're spreading so much crap around this room, Hannah. A month ago that comment would have been out of my mouth in a heartbeat. My relationship with Anna seems to have at least one positive effect. "She has an old acquaintance who is causing her problems. I need you to text me immediately if you notice her doing anything unusual," I answer as honestly as I can without betraying what little trust Anna has put in me.

"Is that why your knuckles are all bruised?" She stares at my hands, suspicious of my cooperative attitude.

"We're discussing Anna, not me."

"I'm well aware we're talking about Anna. Are you? Because almost everything she does causes me concern." Hannah continues staring at my hands.

"Just text me if you notice anything significant, please." *I'm willing to beg if I need to, because you're currently my best option for keeping Anna out of harm's way.*

"Why should I trust you?" She straightens to her full five-four height and tilts her head up to add another half inch. That's when

I notice the tiny white spots painted along her hairline are actually shaped like cranes. Interesting, I store that in the memory bank for future use. "You don't have to trust me. This is about keeping Anna safe."

"I think *you're* the one she's needs to be kept safe from, Joe."

Her accusation bothers me more than I want to admit. "I'd never hurt Anna," I defend myself while straining to hold back my growing frustration. I don't want to mess this up since I need to get on Hannah's good side to have any chance of her assisting me in keeping tabs on Anna.

She fixedly gawks at me for an extended time, but she doesn't challenge me anymore. She must have found some kernel of truth in my assertion. "Won't constantly observing Anna, then immediately texting you cause her to become suspicious?" she asks.

"Her cat's not named Schrödinger, Hannah." Her obstinacy is going to cause me to be late for class. "If you really care about Anna as much as you claim to, this should be a no-brainer decision."

She shifts uncomfortably, still not ready to commit. "Seeing you're only as reliable as a cat when the meat's out of reach, I don't think you're the one I should be contacting, Joe. I think that—"

"I couldn't agree with you more. It's always blackest just before you step on the cat. Anna's in real danger and needs *your* help." If that won't sway her, I don't know what will. Hannah adopted the role of mother hen somewhere along the line in life, and now she's vehemently protecting this particular chick of hers, the one she believes is currently the most vulnerable.

"You'll be getting texts from me continuously," she hedges.

That's as close to a yes as I'm going to get, so I quickly jot my number down on a pink sticky pad laying on her desk. "Here's my cell number, put it in your phone. I've got to get to class." Before she can get a word out, I race from Anna's dorm to mine, hoping Lance has already left. I'm not in the mood, nor do I have time, to explain things to him. *Damn.* He's still here.

"You weren't in the room when I woke up this morning, Joe," he flatly states.

Why does that trivial piece of information feel so incriminating? "You missed me. How sweet, Lance."

"I woke up just before five thirty," he declares as if this is extremely significant.

"So proud of you. Did you do it all by yourself?"

"Stop screwing with me. Where were you?" He's getting quite annoyed. *Huh?* Never thought of it from this perspective, but Lance is the male version of a mother hen.

"That Russian lady that tutors Anna was here asking all kinds of questions. Where you were, what time you left, was Anna with you? Apparently, she disappeared without notifying her tutor. Were you with Anna?" he asks.

Information best left unsaid, Lance.

He notices my bruised knuckles and his expression darkens. "Aaron also came by here looking for you, totally ramped up and making all kinds of accusations. Someone broke into his car and trashed it. He claims it was you."

"Couldn't have happened to a nicer person." Aaron's so fixated on that car it serves him right. I'd never force someone to pay for having a car completely detailed just because of some spilt popcorn. Ever since I found out he did that, I've been harassing him.

"Stop irritating him, Joe. He wants to get you kicked out of school, and now he thinks he's got the proof to do it."

"I didn't go anywhere near his car, Lance." *Don't worry, I'm not going to get arrested, not for that anyway.*

He visibly relaxes upon my assurance that I didn't vandalize Aaron's car. "Fess up then, were you with Anna? Are her shit-hot kisses more than you can handle?

More shit-hot than you can imagine, Lance.

"Did you—" Whatever that next thought was, it got abruptly preempted. "Why are your knuckles bruised?" Before Lance can corner me into admitting things I shouldn't, I head out for class. He's better off not knowing the incriminating details of what I was doing this morning.

III

Part Three

THIRD MOVEMENT IN THE COSMIC SYMPHONY OF LIFE

Mars
The bringer of war and bloodshed

Chapter 30

In 2000, Boris Berezovsky, the original oligarch of the 1990s, fled to England. From there, he continued his campaign of trying to oust Vladimir Putin.

In 2001, Berezovsky was put on Moscow's most-wanted list. He had become to Putin what Trotsky was to Stalin. After a decade of surviving numerous assassination attempts, Berezovsky died under disputed circumstances: Suicide? Murder? The coroner was unable to determine a verdict.

Vladimir Ivanovich Alliluyev, Tatiana's grandfather, becomes the most prominent, outspoken oligarch criticizing the Russian regime.

Tatiana
Monday, October 8 to Wednesday, October 10

My fears continue to chase after me as I rush out of the room to get away from Hannah's and Joe's bickering about cats of all things. Why did I think it would be different in America? My dreams of fleeing my problems are quickly disintegrating; my past has followed me here. Approaching Mudd Hall, I spot Ms. Mizuikova standing guard outside the entrance. To say she's madder than a hornet doesn't begin to cover it. I wish I could leave her there, stewing all day, but that would only create more issues.

When her scan rotates my way, she instantly homes in and advances. "Where have you been? Your father's been trying to reach you about a very urgent matter that ..." While Medusa continues her ranting, all I hear is the hissing coming from the snakes I envision writhing in her hair. *Thank you for putting that image in my head, Joe. Not.* Tuning back into her diatribe, I catch the end of a sentence. "... he gave you that phone expressly because it is critically important he be able to reach you at all times."

Blin. I forgot to deactivate the mute function on the burner phone after we left the post office. Between Joe's one-sided sparring match in the street and the ensuing events on the ride home, I completely blanked it out.

"Call him right now, Anna," her voice drills into me as a few of her wispy, snake-like strands of hair wiggle free.

"You are not in charge of me. My father may give you your running orders, but I make my own decisions." I want to eviscerate her authoritarian, overbearing, pompous attitude. *Ofiyet'!* These bloody emotions are a nightmare. Maybe Joe's right, I'd be better off without them.

Ms. Mizuikova's face momentarily displays uncertainty before regaining its haughty expression. Menacingly looming beside me, she wordlessly relays my father's directive. Contact him immediately. Stepping away from her, I reluctantly slide the phone out of my purse and dutifully call my father. He picks up instantly and demands to know why I haven't been answering.

"I needed to talk to a government official, so I didn't want to be interrupted," I explain in a quiet voice, covering my mouth with my hand for privacy. My feeble attempt at an excuse falls flat, for whenever I'm in a situation that requires impromptu, improvised justifications to my father, I inevitably fail.

"What were you doing in Eau Claire?" he angrily questions.

How could I have overlooked that? Of course my father would put a tracking device on a car he provided for me. "I was preventing Ms. Mizuikova from reading my personal mail," I snap back at him.

"Having anything sent directly to you at Carleton is foolish. Let Ms. Mizuikova do her job." He reprimands me like I'm still a little girl. In his eyes, I don't know if I'll ever reach adult status.

Well, if he's going to treat me like a child then I'll act like one for him. "I don't want her here. I've never wanted her here!" I shout into the phone, purposefully giving up on any attempt at discretion in order for Medusa to overhear me.

"We've been over all of this before," he exasperatedly states. "Only now the situation has gotten worse. You need to accept that."

"Yes, it's gotten worse. The serpent you hired to protect me squiggles after me everywhere, constantly hissing and not giving me a moment to myself."

He lets out a very long sigh. "I made a mistake, Tatiana."

What? He's agreeing with me?

"I irresponsibly let myself become preoccupied, negligent in my duties. by not taking enough time to consider the full implications of that letter you received from Viktor. After further assessment, it seems highly likely that note is fraudulent and connected to a recent incident here. I'm not willing to take any chances with your safely, so I'm making new arrangements for you." His edict is pronounced in indisputable terms. Any hint of a conciliatory tone is gone.

"You can't rule over every facet of my life again, Father. I am not leaving Carleton." I fight back in just as nonnegotiable terms. "There's less risk in Minnesota than back with you or Grandpapa. That letter wasn't mailed directly to me, so no one knows where I am, unless you and your security team heedlessly barge in, leading them to me."

"This is not open for discussion, Tatiana. After your careless visit to the post office this morning, I don't think anyone would need more leading. Remember what happened the last time you took matters into your own hands? You weren't gone more than two days before you were attacked. I'm finalizing your travel details, pack and be ready."

Why is he overreacting so much? I start frantically pacing the sidewalk, already feeling boxed in. And must he keep throwing that incident in my face? "The letter has no connection to that attack at the cafe. I was in the wrong place at the wrong time, that's all."

"That's not all. You need to take these matters more seriously and let Ms. Mizuikova do her job. You're leaving Carleton today." His ominous decree continues stretching across the countries even after the buzz of the disconnected line zings in my ear.

A triumphant-looking Ms. Mizuikova slithers over to start flaunting her newfound powers, taunting that my father gave her strict instructions to accompany me at *all* times. *T'fu!* Storming past her, I enter Mudd Hall for my chemistry class. "You are to start packing immediately," she badgers as she hurries to keep up.

"Get away from me!" I quickly dart into the classroom, knowing she won't physically try to stop me in such a public place. Contrary to my father's orders, I am not packing my things or leaving Carleton, and there's nothing the snake can do about it… at least not while I'm out and about, which I plan to be for the rest of the day.

Furious at my rebellion, Ms. Mizuikova stays coiled around me every second, during chemistry class, during lunch, after lunch and even following me into the bathroom. She's going to drive me insane. When we enter the fine arts building for my music class, she finally stops insisting I have to head to my room and pack. That should raise a red flag for me, except I'm so happy for the temporary relief that I don't really care right now why she's no longer pestering me.

Before long, the initial glow wears off, and her domineering presence glued to my side starts ratcheting up my anxiety. The disheartening fact is that I won't be able to avoid my father's plans for much longer. My angst percolates to the boiling point, causing a volatile concoction of emotions to spill over. This time the recipient is a classmate named Robin Hood. Why would anyone name their child Robin Hood?

"Mrs. Zimmer, isn't it true that Russia has more virtuoso musicians than any other country?" Robin Hood asks. "There's Tchaikovsky, Rubinstein, Stravinsky, Ashkenazy, Rachmaninoff, Horowitz, Richter and now Lugansky. And that's just the pianists." Her ponytail exaggeratedly jumps around with each bob of her head as she haughtily list off the names. "Oh, there's also Van Cliburn," she adds on with even more arrogance in her voice.

"No," I irritably correct.

"Yes, Anna, Van Cliburn is considered a virtuoso musician. He won the Tchaikovsky Competition in Moscow in 1958," she argues.

"No, he eez not Russian." I think I should know. My music education was very thorough.

"*Time* magazine printed that he was the pianist who took Moscow by storm." She says this like it's irrefutable evidence.

Did she read even one sentence of the article? "Texas eez not pardt of Russia. Four downs, yu out, three quarters, game over!" *Yolki-palki!* My effort to learn about that barbaric game of American football is destroying brain cells. I originally thought it might help me communicate more effectively with Joe, but it hasn't turned out that way. "Four penalties for Robeen Hoot, Four Horzes of Apocalypze vil hunt four corners of Sherwoot Forezt du zilenze her!" I rephrase my retort in a context more to my liking.

She stares at me wide-eyed, accompanied by every mouth in the room hanging open with identical looks of shock. Apparently, I got a little carried away venting my anxiety onto an unwitting Robin Hood.

I rush out of the building with Ms. Mizuikova squirming alongside me. When I calm down enough to think straight, a realization hits me—I'm still here at Carleton. I haven't been whisked off to a safe house or the airport yet. What's going on?

My father answers on the first ring. "What's wrong, Tatiana?"

"I'm glad you came to your senses, deciding to let me remain here," I say with a false confidence, fervently hoping it's true.

"Didn't Ms. Mizuikova tell you?" he asks. I glance over to see a sick satisfaction plastered on her face. "I'm waiting until Friday before flying you out so we can mutually agree upon a destination. I don't want you rashly running off on your own again. Until then, I've hired additional security personnel to safeguard you. They should be arriving soon."

More guards! Furious, I throw the phone down and stomp on it. While I'm crushing it to smithereens, I spot the four newly hired bodyguards, or rather *prison guards,* approaching. Is it the sun in my eyes, or is one of them chuckling at my public temper tantrum? *Hmm.* That guard might be a weak link, a possible leak in the dam to chip away at.

The recently anointed warden, King Cobra Ms. Mizuikova, eagerly glides over to greet them with an excited gleam in her eye. She doles out their instructions, then leers back at me, thrilled with these turn of events.

Twenty-four hours of Medusa and her goons constantly infringing on my personal space, not allowing me even an iota of breathing room, pushes me over the edge. I've probably been over the edge for a while now, only this time it's really over the edge. No one should have to live like this, stuck inside a stifling, restrictive straitjacket of protection. I have to find a way out, but with Ms. Mizuikova eyeing my every move, escape appears impossible unless I have an accomplice. And the only person I can turn to for that type of help is Joe.

Don't expect any more help from me. His words ring loud and clear in my memory. Yet, he did turn around and follow me back to my room after our ill-fated expedition to Eau Claire. There must be some way I can convince him to assist me because the rest of my life depends on it. Doing nothing isn't an option, that only leads to nowhere. Hold on—nowhere. That's exactly where I need to be. Using my car won't work, not with my father's tracking device on it, since that car will always be going somewhere.

I impatiently wile away the hours until the wee start of the next morning, one of the few times Ms. Mizuikova and her guards are not hanging around my neck. Slipping out of bed, I boot up Hannah's computer. Fortunate for me, she's careless with her passwords. Logging on, I scour the internet, searching for something suitable. It doesn't take long, *2001 Toyota Corolla, 190,000 miles, dependable, $900.* Perfect. All that remains is to persuade a certain someone to buy it for me.

The rest of the morning I keep alert for an opportunity to talk privately with Joe. As the day progresses, I get gloomier and gloomier, reaching the conclusion that there's little chance of that happening. My father calls while I'm eating lunch, leading to another confrontational conversation. I end up conceding to relocate to a remote location in Wales with the expectation that my agreement will ease up his current security measures, allowing a few holes in the network to open up, one of which I'll break through.

The rest of the afternoon and into early evening I continue my vigilant watch for any small window of opportunity to be alone with Joe. The tide is turning against me. I have to act soon.

Hannah opportunely walks into the room, inadvertently, yet very effectively, blocking Ms. Mizuikova's way. With the buffer she's providing me, I burst out without my human shackles. Exiting the dorm, I charge pass the guards before the snake has a chance to sound the alarm. Sprinting at top speed, I reach Joe's room a full flight of stairs ahead of them. Bolting inside, I slam the door, locking it afterward. *Whew, made it.*

The banging starts almost immediately, followed a short while later by Ms. Mizuikova's demands that I open the door.

I soundly rebuff her through the barrier. "We are four floors up with this being the only entrance to the room. No one can get in or out without you noticing. Your sole reason for existing is to keep me safe, not completely invade every bit of my world!" Turning around, I notice Lance gaping at me, too stunned to say a word. *Wow.* He has gorgeous teeth, so straight and bright. *Focus, Tatiana. You need to stay on task.*

"I'm glad someone is taking precautions for your safety seriously," Joe wryly remarks. We haven't talked for two days, not since our Eau Claire escapade. I wish I could speak with him alone, but if Lance goes out, there's no stopping Ms. Mizuikova from storming in. I cross my fingers, even attempt to cross my toes, employing all possible methods of voiding his previous pronouncement of not helping me, and skirt over next to him.

"I need you to buy a car for me," I whisper. The total look of disbelief on his face isn't a good sign. "I found a suitable one on the internet," I continue.

"What's wrong with your car, Anna?" he asks.

"This one isn't far from here, just over in Farmington. I can get you the money tomorrow, but I need you to buy it immediately." Please don't ask any more questions.

"What's wrong with your car?" he repeats.

"This car isn't far and I can get you the money." *I restate myself when frazzled ... I restate myself when frazzled ... I restate myself when frazzled.* My internal phonograph needle is stuck. I only owned one record growing up, the one my brothers gave me as a prank on my birthday. When I wanted to annoy them, I'd play this track on my parents ancient phonograph player at full blast. Payback's a bear, a mean old grizzly, not a cute, cuddly water bear.

"Anna, what's wrong with your car?" Joe's needle is stuck too.

"Quiet, this has to be just between us. It's extremely urgent and the last thing I'll ever ask of you," I pathetically beg. "I need—"

"No, you don't, not from me. I'm not your *cat's-paw.*" Turning to Lance, he switches to English. "Pull up the definition of 'cat's-paw' for us, Lance, so Anna can hear all of its meanings."

Lance's dark chocolate brown eyes dart back and forth between us. Since Joe and I were talking in Russian, he doesn't know the details of what's going on, yet he has a strange smile on his face. *Ofiyet'!* His teeth are incredibly shiny. Temporarily fixated on him, I detect something's changed since I last saw him. A haircut. It was short before, but now it's extremely closely cropped, pleasantly highlighting his rich ebony skin. Like Joe, Lance is in terrific shape with a cutting, impressive build. Today he's wearing pressed black pants paired with a pale pink button-up shirt to top off his classy appearance. I highly approve of his conservative style, quite a refreshing contrast to Hannah's wild wardrobe.

"The phrase has numerous connotations, some of which are rather obscure, Lance. You'll find them intriguing." Joe tries to lure him in. It works, Lance picks up his cellphone.

"A cat's-paw refers to a hitch knot, a tool for removing nails, an instrument of torture also known as a Spanish Tickler, a species of plant, small ripples on the surface of water or a person unwillingly used to accomplish another person's purpose." Lance slowly recites all the definitions, lowering his voice as he reads the last meaning.

"Thanks, Lance." Joe says, then switches back to Russian. "Just in case you didn't take in all the finer nuances of that definition, Anna, the meaning I intended is the last one. Someone unwillingly used to accomplish the other person's purpose."

My spur-of-the-moment plan isn't functioning. How can I twist things around? "Did you see the sun dogs in the sky this morning, Joe? They were so bright they actually looked like two additional suns."

"Trying to suck up to my love of dogs won't work. I'm not helping you."

"I can't get this car by myself," I frantically plead.

"Fine, Anna, you're right."

Right about what? Joe's so masterful at manipulating conversations, I often get misled. Asking him to explain won't be productive. Desperate to gain his cooperation, I employ my last resort. The truth. "The problem with my car is that my father has a tracking device on it."

"So?"

"So have you forgotten the welcoming party outside your door?" He starts spinning a pen around with his fingers, then switches to drawing demented tic-tac-toe diagrams. "Stop doodling childish games. This is vitally important." I clamp my hand down over his pen.

"My football plays are vitally important too." He pulls the pen back under his control. "Where are you going that you need a different car, Anna?"

"Away from my father, who has unilaterally decided it's too dangerous for me here and is relocating me on Friday." Surely, he can understand the urgency of the situation now.

Crumpling up the piece of paper with his *important* plays, he tosses it into the corner wastebasket. "Then here's my advice for you. Safe travels with your father."

Lightning strikes. He isn't going to help me! How incredibly stupid I've been to think I could gerrymander him into doing my bidding. Lightning strikes again as the magnitude of my miscalculation hits me. I've become dependent upon Joe, in more ways than I want to admit.

Chapter 31

Joe, age sixteen

"Why were you fighting a fellow gang member?" the detective drills me yet again.

These amateur cops drew that conclusion solely on the basis of hearing us both arguing in Russian and can't get past it. Fine with me, let them think that.

"This would go much better for you, son, if you'd answer our questions. You're looking at second-degree-murder charges," another cop threatens.

The door to the interrogation room opens.

"Release him. Mr. Kluge regained consciousness and said that the kid didn't attack him and wasn't part of the break-in. He does works for him around the house."

"Well, it's about time you made it to the hospital to see me," Kluge grumbles.

"I've been a little held up," I bitterly state. And I would have been a lot more held up if Kluge had died. My dad and stepmom are currently in Germany helping my grandfather sell the family home and move into a small apartment, so I didn't bother trying to reach them. Nobody else that has the means to get me out of jail would make the effort to do so.

"I heard that if I hadn't regained consciousness, you would have been charged with murder. Thank you for keeping my secrets, Joseph."

"I didn't keep my mouth shut for you. I had my own reasons," I spit at him.

He eyes me up and down, looking at the stitches on my face and the bandages on my arm. "Now that I know you can keep quiet, it's time I told you more of the story of my life."

Joe
Wednesday evening, October 10 to Thursday, October 11

She's completely shocked that I'm not going to help her. She never expected me to say no. *Really, Anna, what kind of special stupid do you think I am? Did you actually believe I'd jump at your command and race off to purchase a get-away car with your over-protective father's security detail on my heels?* I try my damnedest to purge her from my thoughts, but visions of Anna recklessly charging away alone, completely clueless to what other dangers she might be rushing into, won't quit. She's so damned determined to escape her father's far-reaching hold that she's not thinking straight.

I eventually cave and decide to help her, but not the way she wants; I'm not buying a car for her. Browsing the internet, I search for Elena Proctor's phone number, the woman who forwarded the letters to Anna. *Yes!* She still has that antiquated device, a landline. I enter the number in my phone, hesitating before calling. What if Anna's father is as paranoid and powerful as she claims he is? Contacting this lady could send him spying further into my past. *Screw it.* It's his turn to be pissed off about someone poking around in his private business. I make the call.

"Hallo."

"Hello, is this Elena Proctor?" I ask in Russian.

"Who is this?" she demands in Russian.

"My name's Alek Sokolov. I've been instructed to look into the letters you forwarded to Ms. Mizuikova."

"Who instructed you to do that?" she suspiciously asks.

"Have you noticed any strangers in town or heard anyone asking unusual questions about Tatiana?" I authoritatively question.

"No one until you, and tell Pavel I want it kept that way!" She slams the phone down to disconnect the line.

Another name for the family tree, Pavel. Is Pavel Anna's father? Are Pavel and Elena brother and sister? Could it be that easy? I type *Pavel Alliluyev* into my computer and get one hit, a man assigned to the Soviet trade mission in Germany during the Civil War. This Pavel Alliluyev's youngest sister was the second wife of Joseph Stalin. He was born too long ago though to be Anna's father, but he might be a distant relative of hers. Ties to Stalin could account for Anna's reluctance to talk about herself, but not the current dangers she's facing. I type in *Tatiana Alliluyeva*. Nothing comes up. I put in *Anna Alliluyeva* and get two hits, Stalin's maternal aunt and another woman who died in 1931 without any children.

I'm so sick of dealing with this mess. Lance sends me a questioning frown, disappointed I haven't explained anything about the soap opera drama playing out between Anna and me. Crashing on my bed, I close my eyes, attempting a short mental pause. Of course, images of Anna and Lina accompany me, causing my gut to start churning. After a few rounds of tossing and turning, dreading the dreams that will come in sleep, I decide to get out of here.

Snatching Lance's car keys, I make my way across campus. Just as I'm pulling out of the parking lot, I spy Lina out the side window, frantically waving at me and yelling something I can't make out.

Slamming on the brakes, I force the car to an instant halt. Lina starts running toward me, when out of nowhere, her father appears, shoving her to the ground and viciously kicking her. I have to get out of this car, but I can't get the damn seat belt off! Lina twists her head in a grotesque manner to stare up at me, her eyes filled with betrayal and hurt. I've got to get to her. Swearing at her father, I start pounding on the car window as hard as I can. He looks over at me, laughs, then gives Lina another malicious kick. She's crying uncontrollably now as I wildly pull and pull on the door handle.

"Joe! Joe!" Someone's grabbing my shoulders. I bolt upright for an instant before falling back out of breath. My shirt's soaked through with sweat, clinging to my chest and intensifying the sensation of having the wind knocked out of me. I'm groggy and not sure where I am.

"Are you all right?"

Recognizing Lance's voice, I realize I'm still in my dorm room. "What time is it?"

"Three thirty a.m. Are you all right, Joe?" he repeats.

"No." Rolling out of bed, I brush him aside, change into my gym clothes and put on my almost-glow-in-the-dark running shoes, preferring to battle my waking demons over these subconscious ones.

"Where are you going? It's three thirty in the morning."

"You already told me that, Lance, and I heard you the first time."

"What's going on, Joe? What kind of trouble are you in?"

"Don't feel like talking, Lance." Upon exiting the dorm, I notice movement off to the far side of the building, and I take off jogging at a brisk pace. Glancing over my shoulder, I spot the outline of a man following me. It's too dark to see him clearly, but I'm guessing he's one of Anna's father's hired thugs. He appears to be in good shape, let's see how he enjoys a fifteen-mile run to start off his day. With a grim sense of satisfaction, I pick up my speed. By mile four, he's gone, probably doubled back to the dorm to await my return.

I continue running and running and running, yet no matter how far I go, I can't seem to outrun that damn nightmare. Guilt clings to my every thought, guilt that I'm the cause of Lina's disappearance, guilt that I can't find her, guilt that I kissed Anna, guilt that for the past month I haven't even been trying to find Lina, and now guilt that unless I do something to stop her, Anna will run away too.

Emotionally spent by the time I return to the dorm, I take a long, hot shower, then head for the cafeteria to get some breakfast. That done, I kill time watching football videos. My phone starts ringing, it's Hannah. I don't want to talk to her, but she might have something important to relay concerning Anna. "What do you want, Hannah?"

"In no uncertain terms, you directed me to contact you if there was anything about Anna's behavior that was stranger than normal. So I'm calling."

"No more need to do that, Anna's leaving Carleton tomorrow." I hang up and head to my morning class, where my fears for her safety continue to mount. After class, I force down lunch while debating whether to show up for chemistry lab. In the end, I opt for going, maybe seeing Anna one last time will alleviate some of my copious concerns. Even if it doesn't, I should grit it out instead of ghosting her so she doesn't leave Carleton feeling like the whole world deserted her.

I purposely arrive at Mudd Hall twenty minutes late, due to my lack of confidence in my ability to last through the entire session. I'm only a few steps along the hallway when Anna's tormenting barrage of emotions slam me. Panic, fear and abandonment. *Don't go in there, Joe. There's still one more day until her father plucks her out of here.*

Turning around I retreat, half relieved, half ashamed. The nagging feeling that Anna's in danger spools up its intensity, along with the accompanying compulsion to watch over her. *Leave it, Joe.* Soon she'll be safely sequestered somewhere far away. Until then, she has plenty of people guarding her. Anna doesn't need my protection.

Lina does.

Chapter 32

Tatiana, age thirteen

I've decided I want to become a scientist, an astrophysicist to be exact. I figure if I study the stars, I might be able to determine how and why they sing to me. My first step toward that goal is to enter and win first place at the Russian National Science Festival, NAUKA 0+.

I formulate a step-by-step plan for success, ten steps in all. Implementation runs smoothly until step three. After many days of long, hard thinking, I reach the conclusion that the only thing I possess for a project that would be striking enough for victory are the remaining Silver Bear diamonds. One of them should be sufficient to get the job done, my thirteen-year-old brain logically concludes.

The possibilities of diamonds for scientists are enormous. They're the hardest known material, don't chemically react with other substances, can be tweaked to hold an electrical charge, are fully transparent to many wave lengths and are excellent electrical insulators.

My science project didn't win, but it got me lots of attention. The wrong type of attention. School officials actually interrogated my father about how I got my hands on one of those diamonds. He made up some story about my mother inheriting it from a distant relative who purchased it at a now defunct jewelry store. Naturally, my father figured out Grandpapa had somehow managed to give it to me, even though I refused to admit it.

I'm instantly put into full lock down mode for three months. Those cursed diamonds are two for two at backfiring in my face. Number three will never see the light of day.

Tatiana
Thursday afternoon, October 11 to Saturday, October 12

I have less than twenty-four hours to figure a way out of this house arrest! Deep in the throes of brainstorming, I'm ambushed by Joe's late "arrival" to chemistry lab, although technically he doesn't physically enter the room. After approaching the door, he tucks tail and runs away, but the damage is already done. His powerful feelings of anxiety intensify my already alarming panic. I can't afford to be waylaid by this cursed emotional chain. Determinedly, I fight it, paying a minuscule amount of attention to the experiment and expending the rest on formulating different courses of action for my jailbreak.

"What are you waiting for?" Medusa rudely asks. Looking around, I realize chemistry lab is ending. How did I zone out for all of that? "Well?" she prods me.

"Ded Moroz," I contemptuously answer.

"Grandfather Frost will not be riding his sled full of gifts to visit you this New Year, neither will any other bearer of good things if I have anything to do with it," she grumbles.

The remainder of the day, I continue mulling over my getaway options. There's no way out with every move I make being scrutinized every second. *Think, Tatiana, think.*

After a restless night, I wake with a sick feeling of despair in the pit of my stomach. My father's having me moved today! Nothing else is important anymore. Figuring out how Joe's messing with my head? Doesn't matter. Getting my bachelor's degree? Doesn't matter. Going to chemistry class? Especially doesn't matter.

Ms. Mizuikova's delighted with my morose attitude and highly pleased that I'm silently moping about rather than insisting on going to classes. She was furious when I was able to persuade my father to let me continue attending them.

Bzzzt ... bzzzt. My new burner phone unexpectedly vibrates. "Hallo," I answer with a resigned air of defeat.

"Tatiana, your brother Alex is in the hospital." My father's voice is strained and laced with weariness. "He was standing next to your grandfather when an assassination attempt occurred. He'll be all right, however, I have a few things that need my immediate attention. I'm delaying your transfer until Sunday when I can personally be there to monitor your safety."

Sunday, that gives me two extra days!

"Tatiana, did you hear me?"

"Yes, Father," I obediently reply. "How is Grandpapa?"

"Fine," he bitterly states. "I've had some more reports come in, Tatiana. Keep away from that Carleton boy. He may be involved in all this. He killed a man and—"

"He killed someone?" I exclaim.

"I don't have time to explain. Stay put." He hangs up.

All kinds of questions pop up, along with them, all sorts of tangled feelings join in. I want to ask Joe about this. *Don't go down that road, Tatiana. Best to forget about Joe, you'll probably never see him again.*

"Anna, what was that call about?" Ms. Mizuikova confronts me. As tempting as it is to fabricate something that will send her off on a wild-goose chase, I don't dare risk it. My father will be updating her shortly, therefore, my taradiddle would be foolish. "My brother's in the hospital, so I won't be leaving until Sunday," I mutter.

Back to planning. Saturdays, Saturdays, what happens on Saturdays? Of course! Carleton football games, sporting large crowds in a small area, perfect place to escape. With this broad framework established, my thoughts compulsively revert back to my conversation with my father. *Joe killed someone.* Who, why and what else is he hiding from me? Grabbing my backpack, I head for the door.

"Where are you going?" Ms. Mizuikova stops my progress.

"Chemistry class."

"Attending class is pointless, you're remaining here."

"That's not your decision to make, Ms. Mizuikova. My father said I can continue to attend classes." I race out before he updates her with the entire reason for the delay, along with the additional circumstances why I shouldn't be anywhere near Joe. I'm a third of the way to Mudd Hall when serendipity intervenes, activating the unexplainable link that binds us by a trail of emotional breadcrumbs. I change direction, homing in on where it's coming from, Gould Library.

"What are you doing?" Medusa instantly grills me.

"I forgot something I need for my history paper. I'm going to the library." Why couldn't she have stayed back in my room and let the security guys tag along? They're so much more reasonable.

"Your homework for the following week is totally immaterial. If you're not going to class, we're heading back to your room." Her phone starts ringing. I'm certain it's my father calling to brief her, which will more than likely nix my going to class 'pass.' I motor full tilt, managing to make it into the library. Following Joe's signals through the first-floor study area, I spy him in a chair next to a corner window. He's facing me with his arms crossed in defiance; it seems he was expecting me.

"You killed someone! Were you ever going to mention that to me?"

I angrily accost him, initiating a foreboding air of mistrust to swirl between us.

His expression hardens. "It's not something I talk about, Anna. You should read more Napoleon and take his advice instead of prying into my private affairs. 'The World is not ruined by the wickedness of the wicked, but by the weakness of the good.' "

I feel hurt, betrayed and angry. Soon, I'm so mad I'm seething red fumes, not a good color for humans to detect snakes sneaking up on them.

"You are leaving." Medusa ambushes me, looping her arm through mine and attempting to drag me away.

"Let go of me." I jerk away, backpedaling out of her reach. "I will not be forcibly hauled out of the library like a criminal." The rage in my voice shocks even me. Twisting it into a useful weapon, I round back on Joe. "Who did you kill?"

"Excuse me, miss. If you can't keep your voice down, I'm going to have to ask you to leave."

"Sorri," I lamely mutter. The librarian gives me a stern look then heads back to the research area. I slip into the chair next to Joe, repeating my question through gritted teeth with my 'inside' voice.

"Anna Alkaeva." The cynical way he recites my alias leaves an extremely bitter taste in my mouth. " 'Alkaev,' from the Russian verb 'alkat,' meaning to wish. It's a little too contrived. Did Medusa think that up?" He states *Medusa* like it's a curse word.

She's avidly watching us with gleeful interest, enjoying our spat far too much to end it right away. Joe leans in and whispers in my ear. "Or did you make up that name yourself, *Tatiana*?"

His fireball hits me with a ferocious force. Feeling like my entire world is collapsing in on me, I jump up and slap him as hard as I can across the face. Joe sits there stoically enduring my crazed temper, but I can feel his internal rage. Then the typical pattern of our relationship starts up, the pinging of emotional alphas and omegas. Hostility and compassion, repulsion and attraction, aloofness and intimacy. We both take cover, donning the inevitable Russian cloak of submission to our ill fates, lives dominated by hardship. Why should we have expected anything different?

"In the physics department at the University of Queensland, there's a beaker filled with pitch, a derivative of tar, which originally was thought to be a solid," Joe pedantically lectures me. "It's the world's longest-running lab experiment and was started in 1927 when Professor Parnell heated up the pitch and put it into a glass funnel. During the next eighty years, eight drops fell, proving it wasn't a solid. But no one saw it happen. This experiment teaches us that the world is not always what it appears to be, Anna, and even when you watch very closely, you still may not see it." Joe gives me a cursory nod accompanied by a parting remark. "It was interesting getting to know you."

Just like that he's gone, leaving me without any answer to the circumstances of his murderous actions, only another one of his bloody riddles. I exasperatedly stomp back to my room to blast the kind of music Medusa hates. Making the pretense of laboring over schoolwork, I return to formulating details for my Saturday escape. I'll need a clever ruse to get to the football game and a notable distraction to sneak out of the stadium. I'll also require a disguise, cash and some form of transportation away from Carleton.

Let's see, to alter my appearance, I can dress in a double layer of clothes very early in the morning and remove one layer when I'm in the stadium bathroom. In addition, I'll pull my hair up under a stocking hat and add glasses. For cash, I can withdraw extra money on my debit card at the student store under the pretense of buying a cola. Now on to transportation and destination. These will require a little more work which I need to do away from the nosy eyes of

Ms. Mizuikova. I'll figure something out on the internet late tonight when she's gone, the guards are posted outside the building and Hannah's fast asleep. Good thing she's a relatively sound sleeper.

The final item I need to address is a way to get to and away from the football game. This maneuver will be trickier because unlike attending classes, there's no way I can persuade my father to allow me to do this. I futilely rack my brain for another half hour, then decide to take a break. Picking up my purse, I move toward the door, planning to initiate step one of my scheme. Dropping by the student store to acquire some cash.

As expected, Ms. Mizuikova promptly steps in front of me. "You are not leaving. You're not going anywhere after that stunt you just pulled at the library," she uncompromisingly states, following her script to a T.

We'll see about that. Pulling out my phone, I call my father, over and over and over again. I get sent to his voicemail, over and over and over again, where I vent my copious frustration. The more I call, the more anxious my pacing the floor becomes until it feels like I'm wearing a hole through it. At last on the twenty-second call, he answers and I score a small victory. I convince him that keeping me locked inside this room like a caged animal for the next forty-eight hours is inhumane. Student store, here I come.

Triumphantly, I hand the phone over to Ms. Mizuikova, joyfully watching the results. The longer she listens, the more enraged she becomes, actually starting to dispute my father's decision. While I'd love to keep soaking in the infuriated expression plastered on her face, I take advantage of her inattentiveness to discreetly ready my disguise and write out a brief note. All this is easily completed without her noticing.

When she finally hangs up, I smugly announce, "I'm going to the student store now." She fiercely clings to her remaining power, insisting I wait until all four guards are here to accompany me. The two who had the night shift are sure to be thrilled about this turn of events. Upon their arrival, Ms. Mizuikova hisses out instructions, then announces that she has other more important matters to attend to. *Ypa!* I'll be away from her beady, prying eyes.

We're quite a sight, the five of us trucking across campus. At the student store, I pick out a sandwich, a snack of veggies, a few toiletries, some drinks and a Mars bar. The candy wrapper is black with the word *Mars* written in bright red letters. I can't wait to taste it, even if the flavor isn't that great, how can I not like a candy named Mars? *Don't get distracted, Tatiana.*

The guards, quite in contrast to Ms. Mizuikova's methods, aren't paying super close attention to every tiny, little thing I'm doing. They're too busy scanning the store and surrounding area for possible threats, making it rather simple for me to surreptitiously withdraw extra cash during my purchase.

When we return to my room, Medusa's already back, or maybe she never left. I sit down to thoroughly enjoy the pleasure of noisily crunching my celery and carrots. It was such a boon when I discovered her sensitivity issues with annoying noises.

I'm well in the midst of this joyful activity when Hannah waltzes in, my cue to start conspicuously asking her questions about Carleton football. My intention is twofold—gain more specific information about the game and get Ms. Mizuikova so upset she'll lose her cool.

"How long are de time slotz in foodtball?" I ask.

Hannah is somewhat accustomed to my alternating bouts of silence and sparse conversation by now. "They're called quarters, Anna, and they are fifteen minutes long."

"How long are de breaks in be-tveen quarters?"

"Pretty short except for halftime. That's when the band plays." She furtively glances at Ms. Mizuikova, then whispers to me, "What kind of trouble are you in, Anna?"

"Vat time doez de foodtball game begin on Saturday?" I don't deviate from my plan.

"You are not going to that football game." That didn't take long. Ms. Mizuikova broke quicker than anticipated. Six more questions are still setting on the tip of my tongue.

"Yes, I am!" I switch to Russian. "I contacted my father while I was at the student store and he approved, as long as all the guards go with me." Her face burns bright red while she rants more *you are nots* at me. With a final *harrumph,* she pulls out her phone to confirm my father's permission for my field trip to the stadium.

I secretly slip Hannah the note I wrote earlier. Playing along, she covertly opens it. *Find me in the library tomorrow at 12:45. Bring friends and invite me to the football game.* Instantly understanding that Medusa shouldn't know about it, Hannah pockets the note and slyly winks at me. She's eager and excited to help me, completely the opposite of Joe. The wheels have been set in motion, may I not be crushed within the cogs.

"Your father *did not* agree to you attending that game, Anna," Ms. Mizuikova accusingly calls out my bluff.

"Oh, I guess I must have misunderstood him." Another part of the plan falls perfectly into place. Next step is to covertly get on Hannah's computer tonight to work out my transportation. *Ahhh.* What a cinch it is to pass away the hours when I'm not feeling hopelessly trapped in a snare.

Before I know it, I'm logging on Hannah's computer. There's a bus that leaves Carleton at three and goes to a Minneapolis Megabus stop. Excellent. From there, I'll simply buy the next available ticket out of state that's heading east; I want to be nearer to Aunt Elena. Having that last piece of the puzzle figured out, I settle back into bed with a sense of satisfaction and fall fast sleep.

T'fu! When I wake up, I discover an irate Ms. Mizuikova who has initiated a ridiculous amount of extra restrictions while I was slumbering, probably for the sole purpose of making my life miserable. "You will not be ... and you are not going to ... and you won't ..." The list goes on and on. Somehow, I endure it by continuously telling myself that at three p.m., I'll be riding that bus into the sunset and she'll be eating my dust.

Noontime arrives! I put on my coat and grab my purse.

"You are not going out." Medusa predictably stops me. "You are not doing anything until your father gets here." She's sounding like a broken record, and like a broken record, I'm going to rid myself of her. But after five rounds of verbal sparring with Medusa and six scathing messages left on my father's voicemail, informing him that she is once again keeping me locked in this teeny room, I'm starting to worry. Maybe I shouldn't have cut this so close, however, I didn't want to have dead time at the library. I'm about to start my seventh call to him when my cell rings.

"Tatiana, I don't have time for this."

"Ms. Mizuikova has kept me imprisoned in this room since yesterday afternoon, Father, and she intends to keep me locked up until you arrive. That's over twenty-four hours from now. I'm not a convict, I'm your daughter!" When he starts telling me to calm down, I heatedly yell louder over the top of him, demanding to be allowed to go to the football game. After a few more embattled exchanges, I'm victorious. In order to shut me up and gain my cooperation, he compromises by allowing me to go to the library. And from the library, it's just a short walk to the football stadium. Everything is fitting together so beautifully. I'm really enjoying putting into action the second positive thing I've learned about these emotional outbursts. They are a tremendous aid in managing my father.

Right before we leave my room one of the guards goes down with a bout of diarrhea. Another is looking rather sickly, rushing off to find the bathroom shortly after we enter the library. It might have something to do with the laxatives I slipped into the drinks I shared with them. Good thing I thought to pick those up at the student store.

Promptly at twelve forty-five, Hannah bounces in with six of her friends bouncing along with her. "Anna! Anna! You've got to come to this football game with us," she enthusiastically invites me.

"She is not going to de foodtball game," Ms. Mizuikova commands as an alarming look crosses her face.

"It's the most exciting game of the year. She can't miss this!" Hannah stands right in front of Ms. Mizuikova while her friends surround me. The clever move catches Medusa totally off guard. *You go, girl, Hannah!*

"Anna!" Medusa desperately yells, trying to squeeze herself through Hannah's friends, who are effectively shepherding me out of the library.

"You can't stop me now. Good riddance!" I scream in Russian and wave adieu to her over the top of Hannah's friends' heads. All the ruckus works just as I had hoped and the library staff, remembering us from yesterday, rush over to help usher us out of the building. Once outside, an even larger group of Hannah's friends join the fray, merrily whisking me toward the stadium. Hannah catches up a few moments later and wiggles her way through until she's right next to me.

"How did I do, Anna?"

"Yu are great frend," I admit. Never in a million years did I think I'd be telling her that.

She smiles jubilantly, then shrieks in her squeaky tone, "Seeing the furious look on Medusa's face was better than sliced bread."

In the time it takes to slice Hannah's bread, we're seated inside the stadium. I'm wearing my hat with the wide brim, studying my surroundings and occasionally glancing at football players smashing into one another. Hannah and her friends are on all sides of me. Thankfully, she thought ahead and already had our tickets, unlike the serpent, who had to stand in line and purchase hers. It's not until the start of the second quarter that an irate Ms. Mizuikova and two angry-looking guards finally locate me. Realizing there's no feasible way to remove me from the packed stadium at this moment, they approach and sit as close as possible. Every so often, the serpent hisses a malicious threat in my direction.

Tweeet! A particularly long, sharp whistle blows. All the players stop and mull about. "Eez it de long break now?" I ask Hannah, ready to spring into action.

"No." She chuckles and starts explaining the penalty rules to me. I don't bother listening. I've no desire to expend any more energy understanding this ridiculous game which I'll never watch again. One of her recruits bought popcorn, which I'm impatiently munching on to occupy myself. The middle break can't be that much longer now. Hannah said we'd all go get more snacks then. It's the scripted moment to initiate my jailbreak. With a mixture of anticipation and trepidation, I stare at the game clock, painstakingly watching it slowly tick down. *T'fu.* Will this moronic excuse for a sport ever end?

Out of the blue, blinding pain tears through me. I wrench forward, stick my head between my knees and try to fend off the growing nausea. Hannah doesn't appear to notice my distress and continues to drone on with more banal facts about the game. Medusa doesn't utter a word; she could care less if I'm ill, as long as I'm staying put.

The pain lets up some after a bit, yet the queasy feeling remains, so I keep my head lowered. Hannah gently touches my shoulder. "You can look up now, Anna, it's okay. Joe's walking off the field. I don't think he's hurt too bad." Raising my head, I spy him heading toward the side of the field with a gaggle of people around him. Alarm bells start instantly going off in my head and my hands start to shake. I won't be able to carry out my scheme if this damn connection interferes! I've got to get away immediately, before another blasted episode like this occurs.

"My God, Anna, are you all right? You're deathly white. Say something." Hannah's voice is panicky.

I jump up and rush for the aisle, popcorn scattering around my feet. By some miracle, I manage to race up the stairs just ahead of a mass of students, thus gaining a good-sized lead on Medusa and her gang. Running to the nearest bathroom which I scoped that out on the way in, I slip into a stall, fling the door closed and gulp in some deep breaths.

Quickly, Tatiana, quickly. As I'm frenziedly taking off my sweatshirt, a hand appears from under the stall next to me, pushing over a pile of clothes with a note attached. It's written in Russian. *Your brother will be killed unless you follow my instructions. Put these on.*

In any other circumstances, I'd never submit to blackmail, but with Medusa close on my heels my choices are limited. Creating a scene would waste valuable time and draw unwanted attention. I don the faded, flowered house dress, moss-green knitted sweater, old creased black shoes, gray wig, wire spectacles and large-brimmed straw hat that were shoved at me. This could be an advantageous change of course, for this disguise is far better than mine.

When I open the stall door, a gloved hand firmly seizes my arm and steers me out the exit. "Hunch over and act like an old woman," the person softly grunts in Russian, using a voice that could be either masculine or feminine. Keeping our heads lowered, we weave through the very large crowd with me struggling to maintain a fast pace in shoes that are slightly too big. Occasionally, I sneak glimpses of my captor, taking in as many features as I can. The person is about my height, slender but wiry, wearing battered pink Nike sneakers, wrinkled gray slacks, a baggy pink knitted sweater pulled up around the neck, black sunglasses, a stringy gray wig that covers the sides of the face and a large floppy hat that shields most of the remainder of the face. Nothing gives me a clue as to who this is.

We bump up the speed after exiting the stadium, even though it's completely out of character for an old, bent-over woman. I don't care. The sooner I can distance myself from this place, the better. My captor suddenly slows down to scan the immediate area. No one is close by. Spotting a mostly concealed area, he or she—I'm not completely sure yet but leaning toward he—pulls me in, rapidly removing the hat, wig and sunglasses while I study his appearance. Closely cut brown hair, hazel eyes, narrow nose, cleft chin and a five-centimeter scar along his right cheek. Before I can make more mental notes, he grabs the hat I'm wearing and shoves the wig at me. "Take off the sweater, give it to me and put on this wig."

I briefly contemplate fleeing, but I'm still too close to the stadium to take the risk of drawing attention. I do as he says, then he hides all the items under a bush and yanks me along toward Willis Hall. *Yes!* Just the direction I want to go. We're almost to the bus stop when he abruptly halts us. "Give me your phone."

"No. I'm not—"

He rips my purse from my grip and I viciously fight to snatch it back. He swipes my feet out from under me and I fall sideways on the grass. As I scramble to get up, he's already pulled my phone out of my purse. "Unlock it, or your brother will suffer the consequences."

"I'm not my brother's keeper and you—"

Roughly clutching the collar of the dress I'm wearing, he twists it to raise me up a centimeter by my neck. "We don't have time for this. Unlock it, now." The immediacy of the situation re-registers with me; arguing will consume precious minutes. Grabbing the phone, I unlock it.

"Call Joe's number." He releases me and waits expectantly. The second I punch it in, he snatches the phone back. I hear the call go to voicemail. "Joe, I got the package," my captor triumphantly states after the beep.

What!

Chapter 33

Joe, age sixteen
Two weeks after Kluge is released from the hospital

"Lina is your granddaughter?"

"You're not very bright if you haven't figured that out after all these years," Kluge snidely remarks.

"How the hell was I supposed to figure that out?"

"All the hundreds of questions I've asked about her. Think, Joseph."

"You've asked a lot of questions about a lot of things." The damn bastard, why didn't he ever tell me this?

"You still don't see it, do you? I purposely went to Pastor Gordon and specifically asked for your help so I could have a connection to Ellie, yet remain distant. There's no other reason I would have tolerated all the bullshit you've put me through."

He's been using me all this time. "Screw you! I'm telling her about this and you can go fuck yourself."

"You'll be signing her death sentence if you do that," Kluge says in an ominous tone.

"There's no one out there hunting for you and your precious Silver Bear diamonds. It was half a world away and over half a century ago. Nobody cares if you're alive or dead anymore. You're just a paranoid old man."

There's momentary surprise in his eyes. He didn't know that I'd figured out he had some of those diamonds. "You're very perceptive, Joseph. Then tell me, who was that Russian man you killed at my house?"

"The police linked him to a drug gang. It had nothing to do with you."

"Do you know that or is that what you think? I'm not willing to risk Ellie's life to prove it, are you?"

Joe
Saturday afternoon, October 13

The instant I'm slammed to the ground, intense pain tears through me. When I try to get up to see if we scored, an alarming dizziness and nausea hit me. Falling back to my knees, I look at my hand. It's messed up pretty bad, at least it's not my throwing hand. Weirdly, I'm having a hard time breathing. *WTF?* I've been hurt way worse than this before and never passed out.

The trainer approaches and kneels beside me. His questions help deflect the bizarre anxiety I'm experiencing. By the time I get up to walk off the field, I feel more like myself. *Shit!* Another wave of nausea hits. Where the heck is this coming from?

Anna, it must be Anna. But what's she doing here? Her father was supposed to whisk her away yesterday. Before I have a chance to peruse the crowd to see if I'm right about her still being at Carleton, I'm ushered out of the stadium. The anxiety starts to recede. When I'm in the car, being driven to the emergency room, it fades into background noise.

Three hours later, I'm still stuck in an exam room of the hospital, waiting for the doctor to return. Finally, she bustles back in with a bunch of bandages and gets to work. I try to convince her I only need an Ace bandage, advice she completely ignores by immobilizing my hand and wrist with a splint that goes all the way up to the elbow.

"You are probably going to need surgery. Return in a few days when the swelling goes down for reevaluation," she instructs, then washes up at the sink before moving on to the next patient.

Just great. No way I can practice football wrapped up like this. The assistant coach who came with me to the hospital walks over to greet me as I enter the waiting room. I relay what the doctor said, and he gives me a ride back to campus, dropping me off near my dorm. Wanting to be alone for a while, I wander over to the arboretum, looking ridiculous out here in my football uniform, but I don't give a damn. After aimlessly strolling around for a bit, I end up on the hill overlooking Stewsie Island.

An overcast sky is masking the moon, just like the night Anna and I sat here. The association jogs my memory and her words begin playing back at me. *Because of you, I can't keep this latch in my mind closed that locks out emotions! Whenever you're near me, I have to battle a constant onslaught of feelings that I've spent most of my life without. It's totally excruciating and totally incapacitating!*

Why do I have to remember everything? If only I could purge some of the archives that keep torturing me. *I stopped feeling the joys, sorrows, pains and loves of life. I stopped being human.* "Just what do you expect me to do about it, Anna?" I shout at the night sky.

Joe, help me. What? Where did that come from? I'm not sure if I'm remembering this or if I actually heard it. Cursing, I look out over the water, only to get my own words battering me. *I'm drawn to you too, Anna, and I don't want to be. No matter how much I try to stay away from you, I can't. Because I'm haunted by this feeling that you're in danger.* Vigorously I shake my head, trying to get rid of these flashbacks. This hill overlooking Stewsie Island is obviously the wrong place for me to be.

I walk toward the main part of campus with the vivid memories strutting right along with me. *I'm beginning to accept that the grief will always be there. I'll never stop grieving for her, because I'll never stop loving her. The two are interwoven.*

The two are interwoven, Lina and Anna! I trod faster, but my conscience keeps pace. *He hit her again. I couldn't uselessly stand by and do nothing. So Lina did the only thing she could think of to stop me. She disappeared.* Heartache rips through me for the billionth time. *Shit!* I can't do this to myself again. *Facts, Joe, think about mundane facts. Don't let the memories control you.*

Fact: On October 6, 1995, astronomers discovered the first planet orbiting another star in our galaxy, 51 Pegasi b. *Different facts, Joe.* Fact: On February 14, 1990, Voyager 1 was roughly 3.7 billion miles away when its camera turned for one last glance. In that final photo, Earth was just a dot of light, 0.12 of a pixel in a frame of 640,000 pixels. It—

"Where are you going?" the snaked hisses at me from behind.

Where did she come from? Clearly, I haven't been paying any attention to where I'm going or I wouldn't be anywhere near her. "None of your damn business, Medusa."

"What did Anna say when she called you?" she demands. The strange anguish in her voice, an emotion she hasn't displayed before in my presence, causes me to pause. Anna called me? *Don't get suckered in, Joe.*

Medusa circles menacingly around. "What did she tell you?" The viper is hissing louder now.

"She said she's afraid of snakes," I answer and brush past her. Why is Anna still here? And why is Medusa interrogating me about her?

"I'll ensure you spend the rest of your days in purgatory, ruing the moment you were born. You will pay for this in blood!"

Pay for this in blood? Suddenly, the significance of her scorching threats hits me full-on, singeing the hairs on the back of my neck. Anna succeeded in her escape plan. That urgent, impelling need to protect her becomes so strong that I actually resort to cooperating with Medusa. "I don't know what Anna said. I was brought to the hospital right after I got hurt and don't have my phone. What kind of trouble is she in?"

"None of your damn business," she smugly spits at me, then dementedly slithers off. Crap, I shouldn't have pissed her off so much.

Help me, Joe!

Is that my memory talking to me again or is Anna somehow sending me a message through this eerie connection of ours? A chilling sense of foreboding sets in.

Why did you have to run away, Anna?

Chapter 34

Once I knew only darkness and stillness … my life was without past or future … but a little word from the fingers of another fell into my hand that clutched at emptiness, and my heart leaped to the rapture of living.

Helen Keller

Tatiana
Saturday afternoon, October 13

"Joe, I got the package," my captor states in Russian. He ends the call, then tosses my phone and purse into a nearby garbage can.

"Aren't you concerned that your fingerprints are on my phone?" I query him.

"I'll be long gone before anyone puts all the pieces together, Tatiana. Aren't you at all concerned that your boyfriend is part of this?" he counters.

No. I don't think Joe has anything to do with this. If he wanted to kidnap me, he had numerous opportunities before now which were far less complicated and risky. Also, Joe had no idea my departure was delayed. Even though I'm certain he's hiding things from me, a kidnapping plot is not one of them.

Jerking my arm, my captor rushes us toward the bus. His timing is impeccable and we board the Northfield Metro Express moments before it's due to depart, the very same bus I was planning on taking. This couldn't have worked out better. He prods me to walk two-thirds of the way toward the back, where there are plenty of empty spots, then shoves me into the seats on the left side. Defiantly, I plant myself in the aisle seat; I don't want to be boxed in.

"Do you really want to cause a scene here at the campus bus stop, within the reach of Ms. Mizuikova's iron grasp?" he haughtily whispers.

No. Annoyed, I slide over to the window seat and he settles in next to me. It seems he's done a thorough job with his "spying on Tatiana" homework, capabilities that shouldn't be underestimated. It's best to wait until reaching the Minneapolis Megabus stop before I make my move.

As the bus pulls out, I begin a more thorough examination of my captor. There's a bit of sweat on his brow and upper lip from the power walk we did to get here. To cool off, he takes off his gloves and pushes his sleeves up. A roaring-tiger tattoo is on the inside of his left forearm. That's not good. The roaring-tiger is common among Russian gang members who have committed a murder. Is he still part of that gang? I sure hope not, because it will be far more difficult for me to escape if he's working with others. Time to get him talking, maybe he'll slip up and tell me something useful. "Aren't you worried that I'll be able to identify you?"

"No. Aren't you worried your boyfriend's a part of this?"

We already covered that topic. "Once we're far enough away from campus, I will no longer have need of your services and will be getting off this bus alone." If he tries to stop me, I'll kick, yell, fight and stab him if I have to. Someone is sure to step in to aid a 'helpless' woman.

A wicked smile slowly crosses his face. "That's not the way this story goes. I have quite a different role for you, Tatiana."

I'm not into role playing. "How long were you in the military?" I ask. Since a one-year conscription is mandatory for Russian men, this topic is as good as any to take advantage of the fact that he appears willing to run his mouth off.

"Long enough to acquire many useful skills," he says without hesitation.

"Like what?"

"Like how to manipulate people. This is the first day of your new life, Tatiana. Who you were yesterday doesn't concern anyone anymore." An evil sparkle gleams in his eyes, intensifying the wicked grin still lingering on his lips. Reaching over, he grabs my chin with his thumb and forefinger. "What do you really know about yourself? What are you capable of?" He's quoting a Russian military recruitment video, but aiming the words at me—at Tatiana. I briskly swat his hand away.

"Not what you were expecting? I could just as easily have told you I support the dedovshchina, that brutal practice of violent hazing that instills terror in Russian soldiers. However, I thought that would be a little too strong-armed so early in our relationship."

So instead of tormented fresh recruits, you terrorize young women. You chose the wrong one this time. "How much ransom do you want?" I ask, not swayed by his intimidation tactics.

"What makes you think I want money?"

"Because you're not as smart as you think you are." He certainly didn't go the extra mile in his planning and do a full bio on my family, or he'd never have come anywhere near me.

"I'm smart enough to capture you." A sick laugh rumbles from deep in his throat.

"That wasn't a wise move. You won't live longer than a few days," I warn him, having full faith in my father's and grandfather's abilities to track him down.

"I thought you were a realist." He mockingly cluck-clucks his disappointment.

"I am. My father will hunt you down, and he won't stop there," I coldly inform him.

Another deep, unnerving chuckle erupts from him. "The world has been cruel to you, just as it's been cruel to me, except I have already accepted what you haven't. In the end it doesn't matter, the human race exists in obscurity. There's no interconnectedness. We're all alone ... *You're* all alone." He tauntingly places his hand over his heart. " 'Even though I live in the midst of people, I'm always alone.' "

Anyone who spent any time observing me could deduce that. Yet he's quoting my exact words, ones I spoke to Joe on the trip to Eau Claire.

" 'The woman who follows the crowd will usually go no further than the crowd. The woman who walks alone is likely to find herself in places no one has ever been before.' How very fitting for your current situation, don't you think?"

Now he's parroting Joe, or rather, he's parroting Joe's parroting of Einstein. How has he been listening in on my conversations? *Keep him talking, Tatiana. All his foolish yammering is sure to yield something helpful.* "I think you should be more concerned with *your* current situation. You have a very unrealistic view of it."

"Oh, Tatiana, don't brush aside what you know to be true. Unity doesn't connect people, it destroys them. Seeking unity is an absurd game the world plays, and it's one that you and I have completely failed at. I, though, see it's a hoax and no longer participate."

"Then stop participating in this 'game.' Let me ago and perhaps you'll live to see next week. Although maybe not, you're far too quixotic to succeed as a lowly gang member. Is that why you fled here, to avoid certain death from fellow members who couldn't tolerate your philosophizing any longer?"

"Ah, the quandary you put me in. Do I give up because of the futility, or do I go elsewhere to escape my fate?" He tilts his head and knowingly winks at me. "London perhaps, or maybe Vermont."

Enough fruitless banter. "Why did you send me letters signed with Viktor's initials?" I grill him.

"Viktor, what a fool," he scoffs.

Thanks for confirming you knew Viktor. Let's see if you'll make any more mistakes. "Where did you meet him?"

He gives me a sly look, aware of what I'm up to. "Viktor told me to pass along a message to you. 'I'll come back.' What a ludicrous thing for a dying man to say."

Funny. That's just what I'm thinking about you right now. "Viktor was killed more than three years ago. Why did you wait so long to kidnap me?"

"Isn't it strange how odd connections send us in a direction we didn't intend to go? Take Viktor, for instance. He became quite buddy-buddy with me, whom he claimed he despised, once he was shot and dying. He told me all about his love for you, about your rich grandfather, and your favorite aunt in the United States. So sad *this* is the lasting legacy he gifted you." He drums his hands on his leg, then strikes an imaginary cymbal, wanting a knee-jerk reaction from me.

I won't give him the satisfaction. The bus abruptly jolts to a stop. Both of us become hyperalert and scan for threats. False alarm, a routine stop, the bus driver just hit the brakes too hard. Two young guys in their mid-twenties board the bus, proceed toward the back and sit down a row in front of us. They're both tall, strong and appear to be hotheaded, two ideal bystanders that should be willing and capable of restraining this "Viktor" if he gives me any trouble when I want to get off.

"Tatiana." My captor taps my shoulder to regain my full attention.

"Do you want to know how I've been eavesdropping on you?" He's so eager to show off his prowess that he doesn't wait for my reply. "While Joe was beating the crap out of the wrong guy outside the Eau Claire post office, I slipped a listening device into your purse. The Soundbug, the smallest spy GSM bug in the world," he proudly states.

I've been getting far too sloppy. That's got to stop, here and now. "It doesn't matter what you've heard. You're still a dead man," I bluntly state.

"Quite the contrary, the intriguing tidbits I've heard have been extremely useful, like getting your boyfriend involved in this kidnapping, using the threat on your brother's life to my advantage and learning how to manipulate you."

"No von controlz me!" I purposely yell in English. One of the guys seated in front of us turns his head to stare at us for a few seconds. Excellent, the seed is planted.

"That was very clever, but it won't stop my plans. Enough chatting, we're getting off at the next stop."

"I'm not going anywhere with you." Even though he threw my purse away, I still have some money tucked away in my socks, as well as my ever-present switchblade. We're far enough away from Carleton now that causing a public spectacle is inconsequential to me. By the time Ms. Mizuikova and her posse get wind of it, I'll be long gone. Staring out the window, I discreetly start to reach for my knife.

Ahhh! There's a sharp pain in my thigh.

"That's etorphine M99, a synthetic opioid." There's an added hint of excitement in my captor's voice. "Since I diluted it and administered it intramuscularly, you won't be fully unconsciousness for another fifteen minutes or so. It has some rather nasty side effects though, increased blood pressure, increased heart rate and possibly death.

Fortunate for you, I have the antidote. So, Tatiana, you will be getting off with me."

The bus pulls to a stop across from the IDS Center and my captor grabs my hand. I reluctantly follow him up the aisle, off the bus and onto the street. *Escape now!* I don't dare. I'm already feeling a little dizzy and I can't end up in a hospital. My father could effortlessly track me down there.

After flagging a taxi, my captor roughly shoves me into the back of it. No sooner am I seated, then I start feeling faint. *Concentrate, Tatiana, you need to pay attention to where he's taking you.* My heart begins pounding loudly. Is it my imagination, or is this drug working faster than he said it would? I attempt to take mental snapshots of the buildings we pass, intermittently closing my eyes to avoid passing out. Shortly, the cab pulls over and the driver states the fare. My vision is narrowing. *Joe, help me!* I internally scream my emotional plea, hoping it will reach him.

"Don't give me any trouble," my captor whispers in my ear. Like I have a snowball's chance in sunshine of doing that. Walking around to my side of the taxi, he opens the door, reaches in and grabs me beneath my shoulders. Once I'm out of the cab, he puts his arms around me like we're lovers. "She hadt a little du much du drink," he tells the cab driver as he shoves some money at him.

Half dragging me along and going rather briskly, considering my condition, he moves us down the sidewalk and around a corner. Leaning me against a brick building, he quickly picks the lock on a side door, then we head inside, follow a short hallway and enter a stairwell. By now I'm barely functioning, there's no way I'm navigating stairs, even with his support. He picks me up and painstakingly steps downward.

I pinpoint my total attention on breathing. At the bottom of the stairs, he again props me against a wall. Through the invading fogginess, I see him unlocking a padlock and pushing open a metal door. He brings me in, lays me on the cold concrete floor and kneels next to me to inject what I hope is the antidote into my thigh. I'm incapable of any resistance. My arms get tied behind my back, my feet bound together and a gag is shoved in my mouth.

"People have the unrealistic expectation that life should be pain-free. Pain is as much a part of life as death, Tatiana." His footsteps recede, the door closes, the lock clicks and darkness fills the room. A moment later, the tiny sliver of illumination that was slicing under the door disappears. Complete, utter darkness reigns.

Chapter 35

This is the distinctive turmoil of Einstein's life. He was a self-professed lone sojourner; his thoughts and spirit communed with the cosmos. Yet he could not escape from the often-horrendous acts of humanity.

Joe

Saturday night, October 13

After Medusa slithers off, I decide to visit Hannah to see if she knows anything. Apparently not all of the brood followed their matriarch, because I notice one of the viper's spawns behind me. Anna's father must have some incredibly far-reaching influence if he was able to persuade school officials to allow these guys to roam campus like they do. Or maybe he didn't ask anyone's permission, it's just no one has complained about them yet. Well, I'll fix that. *Damn.* I don't have my phone. That problem will have to wait.

I bound up the stairs to Anna's and Hannah's room, leaving the viper's spawn waiting outside in the shadows.

"Get out of my room," Hannah immediately demands when I walk through the open door.

"Where's Anna?" I quietly ask.

"Get. Out. Now!" Her high squeal pierces my ears.

"Just tell me where she is and I'll leave." She must know something.

"I'm not telling you anything."

Glancing over at Anna's desk, I see it's cluttered with crap. Empty food wrappers, pop cans, notebooks and pencils. That's not like her at all. "I'm sorry I didn't listen to you yesterday, Hannah. Please, tell me where she is."

"Anna's not here," she snipes. "Now get out."

"Do you have any idea where she went?" Desperation is filling my voice and probably written all over my face as well.

"She told me what you did." Hannah's looking at me with pure malice.

Did Anna tell her I killed someone? There's no way Hannah will help me if that's what happened. "What are you talking about?"

"Medusa told me everything. You're more of a snake in the grass than she is."

"I'd never do anything to hurt Anna," I firmly and earnestly state.

"Don't piss on my leg, then tell me it's raining. You're lying like a no-legged dog." Giving me a look of utter disgust, she dramatically points a menacing finger at me. "You assaulted some girl and that's why you got expelled from high school."

The serpent certainly knows how to rile people up, I'll give her that. "I'd never lay a hand on any woman, and I'd be in jail now if I did. Medusa is filling your head with lies. She's not Anna's translator, she's just a lackey hired to keep watch over her."

"Why does Anna need all that security anyway?" she snaps at me.

"I'm not sure, but something's wrong and Medusa's only using me as a scapegoat, someone to blame for her screw-up." I edge over to Anna's desk and peek at one of the papers on top. There are numerous sketches on it, rudimentary cat figures, some starfish-shaped stars and a colored rendering of the planet Mars, but nothing telling of where she might have gone to.

"Well, she's doing a fine job of blaming you," Hannah coldly states as she walks over, probably to make sure I don't touch any of Anna's things.

"You actually believe Medusa more than me?" I ask incredulously.

"Why did you get expelled from high school?" she snipes at me.

"I got caught with copies of some final exams, and that's more truth than you'll ever get out of the viper." I look Hannah right in the eyes. "I would never hurt Anna."

She turns away and starts absentmindedly cleaning the garbage off of Anna's desk, crunching it up and sticking it in the trash. "I don't know where she is, Joe. Anna never tells me anything."

Good, she's coming out from under the viper's spell. "When's the last time you saw her?"

Hannah stops tidying up and heaves a heavy sigh. "At the football game."

"Where was she headed when she left?" There must be some details she can tell me.

"I don't know."

"Exactly where in the stadium were you, and what time was it when you last saw her?"

Hannah's getting annoyed by all my questions, but she still answers. "Up in the stands near midfield, right before halftime. When you got hurt, she turned as white as a ghost. I thought she was going to puke at any moment. Then she just bolted, leaving Medusa spinning in her wake. That's all I know."

"Thanks, Hannah. I really appreciate it."

She gives me a meek smile. "I'm worried about her. Let me know if you find out anything."

I nod my head and take off. Where would Anna go? I have to check my phone and see what that message says. Hopefully, Lance grabbed all my stuff from the locker room. Medusa's minion follows me back to my dorm, then sets up sentinel duty nearby when I enter.

"How's your hand?" Lance asks as I walk in.

"Messed up. Did you get my stuff from the stadium?"

"Yeah, it's on your bed."

I shuffle through it, find my phone and listen to my voicemails. "Joe, call me as soon as you can." *Stop worrying about me, Beth.* I delete her message and listen to the next one. "Joe, I got the package." Who the hell is that? There are no other messages, so that has to be the one Medusa claims was from Anna.

"Do you need to get it operated on?" Lance asks.

"I have to go back in a few days to get it rechecked." *Enough questions, Lance. I've got more pressing issues to deal with.*

"Ms. Mizuikova paid me a visit tonight. She barged in here full of some very damning accusations against you. Are you involved with Anna's disappearance?"

Not him too. "No. What did she tell you?"

"That I should keep better company, that you raped some girl, murdered some guy and were thrown out of high school for selling exams in order to buy drugs." He shrugs it off. "I told her kidnapping and murder weren't on your list of illegal activities and she needed to look elsewhere." He pauses uncomfortably. Something else is gnawing at him. "Did you help Anna get away, Joe?"

"You're too kind to that viper, Lance. You should've told her to go to hell." Seeing the growing alarm on his face, I answer his question. "No, I didn't help Anna get away."

Visibly relieved, he starts to ramble on about his newest pet subject. "Gravity is now thought to exist in a fifth dimension. Isn't that fascinating?"

"Yeah," I mutter and turn on my computer. Not that I expected him to, but Lance doesn't take the hint, he just keeps droning on and on. Usually I can tune him out, but with everything that's going on in my head right now so dramatically out of whack, I'm failing.

"In Einstein's mathematics, space has three dimensions and time is the fourth. Yet according to Einstein's theory of general relativity, gravity acts to bend space-time. Presently, gravity is being looked upon as existing in another dimension, a fifth dimension. What are your thoughts, Joe?"

"Not what I need to be focusing on, Lance." I start looking up bus routes that depart from campus at about the time Anna took off. She wouldn't have used her car, not with her father tracking it, and I highly doubt she found anyone else to buy that Toyota for her.

Lance forges ahead, ignoring my lack of interest in discussing gravitational theories. "To find out what geometry might emerge in the fifth dimension, Sonner used holographic duality to map entangled quarks. Guess what he found?" That excited warble that occurs whenever he gets really keyed up has entered his voice. Maybe if I answer him, he'll shut him up.

"Wormholes, Lance. Einstein and Rosen mathematically showed that back in 1935, it's old news."

"Exactly, Joe, wormholes! The entangled quarks created a wormhole. Therefore, it may not be gravity that holds the universe together, but entangled quarks. I wonder if Einstein ever considered something like that when he wrestled with his concept of *spooky action at a distance*."

From all the multitude of scientific subjects being studied, Lance had to single out this one. I don't need more spookiness and I sure don't need more entanglements. In the same way that Einstein was haunted by quantum entanglement until his dying day, I'll be forever haunted by my spooky entanglement with Anna.

"Do you think he took that into consideration, Joe?" Lance asks again.

"I just told you he did," I irritably reiterate.

"No, you're not listening to me. The Einstein-Rosen wormholes predicted the connection of black holes, not entangled particles. This theory is something new."

Black holes. You got that right, Lance. Anna and I are two entangled black holes.

Thinking he hasn't made his point clear, Lance walks over, places both his hands on my desk and leans in to obstruct my view of the computer screen. "Think of the possibilities, Joe, both entangled particles and wormholes imply faster-than-light travel across the universe."

"Can you just shut up, Lance?" I glare at him and he finally backs off.

Ping. An email comes in; it's from Viktor.Paskhin@gmail.com. *Joe, we got the package.* Another incriminating message from a dead man. Who's trying to frame me now and for what?

Sing for her. The words shoot through me. *"Nothing sings like the stars, nothing shines like the star thrower."*

"Damn it, Lance!" He's connected his iPod to his Bluetooth speakers. He must be really torqued at me, because he knows I hate listening to music.

"She's lost in the darkness, with nothing to protect her. Search the stars to find her footsteps."

Why did he pick this song from the hundreds of songs on his iPod? *Crap!* I should have helped Anna with her outlandish scheme, at least then I'd know where she is. "Put on your damn headphones, Lance!"

"Yet you let her go. Now nothing torments like the memories." These lyrics were written purely to torture me. Abruptly, an image of Lina materializes, then morphs into an image of Anna. I'm on the edge of two worlds, without clear reference of where I am, let alone where they might be.

"Sing for her."

Sing what? What does that mean? As the lyrics continue goading me, both Lina's and Anna's faces appear at once, then meld together to create a distorted, blended apparition that begs me for help. What the heck? My memory never does that! I always remember everything exactly.

"For God's sake, turn off the music, Lance."

"Just let it go, Joe."

I can't let it go, Lance, because I remember everything. I remember the last time you used that phrase, 'let it go.' It was September twenty-seventh 11:36 a.m., after my first Chemistry lab with Anna. Do you remember? Of course not. Only I'm cursed to remember every single moment of every single hour of every single day of my existence.

Chapter 36

Eternity is not endless time

Eternity is when time no longer exists

Gazing in amazement at the cold yet profoundly moving beauty of the eternal, the unfathomable, life and death flow into one, and there is neither evolution nor destiny, only being.
Albert Einstein

Tatiana
Late Saturday night, October 13

NLP. No light perception, absolute blackness, total enveloping darkness. This is what I wake up to.

Wait a minute! That stupid wig is still covering my eyes. After a few contortions, I manage to push it aside, but it doesn't make any difference, it's still pitch-black. I feel like I've been shoved through the event horizon of a black hole; my sense of time is drastically warped, I have no contact with the outside world and I'm trapped on the other side of a space-time boundary with no idea how to get out. This isolation is stifling and frightening.

I need to familiarize myself with my surroundings. Scanning the area, I strain to see anything, yet absolutely nothing is visible. I have no idea which way the door is, so I choose a direction and start shimmying. Almost immediately, I bump into a cement wall. With extreme difficulty and many contortions, I manage to get myself into an upright seated position. That was exhausting. Once my breathing slows, I scooch along the wall. My legs are getting cramped, my hands are numb, my back aches and my head's pounding, still I keep going until I end up bumping into what I think is a large cabinet. The effort it took to go this far further irritates the soreness where my hands and legs are bound. Also I'm quite winded because I can't get this gag out of my mouth. Nonetheless, I'm not giving up.

I wiggle my way around the cabinet and proceed another few meters or so where I reach an adjoining wall. Out of habit, I close my eyes to concentrate. *This dark*—No! No phantom of the opera lyrics. I need to figure of a way out of here. Yet, who is my phantom, the man who captured and sequestered me in this dark, dank place? Is he a member of the Solntsevskaya Bratva, Russia's largest organized crime group? Or is he working with the Russian regime to get at my grandpapa? Maybe he's operating solo or with a partner purely for ransom money. Most important for the immediate future, where did he go and when is he coming back?

As if I summoned him, the lock opens and a shaft of light shoots in, silhouetting a lone figure. The door is quickly closed and a small penlight is turned on. "Robho, it's dark in here. Where are you, Tatiana?" My captor searches the area where he left me, then starts sweeping larger arcs with his flashlight until he finds me crouching in the far corner. Strutting over, he leans down to remove the gag.

I instantly scream at the top of my lungs, earning myself a brutal strike across the face. "Do that again, and I'll inject you with more M99," he angrily snaps.

I can't have that. It's essential I stay conscious. "What do you want with me?" I snip at him while furtively perusing the small area of vision his penlight affords. Stacks of sagging and bowed-out cardboard boxes are precariously perched nearby. Phew, dodged a bullet there, one good bump and those would have toppled over on me.

"You already figured that out, Tatiana, money. Very soon now, Joe's dear friend Aaron will deliver the ransom note to Ms. Mizuikova, then Joe and I will get a very big paycheck." He rubs his fingers together, itching to get his hands on it.

Aaron? The Aaron who blamed Joe for vandalizing his car and trying to kill him in Chemistry lab? I don't think so, he hates Joe's guts. "You don't lie very well. Aaron would never help Joe with anything." Also, while Aaron's a complete twit, he's not an idiot. He wouldn't get involved in this reckless kidnapping scheme. And neither would Joe. "Joe is not involved in your foolish plot either. You're the only one that's dimwitted enough to attempt this."

A hearty laugh explodes out of his mouth and nose. "Still feisty, aren't you? Yet, the isolation is starting to affect your reasoning skills because Aaron did help me. He's so enraged at Joe that he eagerly agreed to deliver a damning love letter that should destroy Joe's romantic relationship with you. Would you like me to read it to you? It's a work of genius. I was able to made it sound tragically poetic while cleverly incorporating my ransom demands."

"Joe and I don't have a romantic relationship." I set him straight.

"You really don't see it, do you?" He shines the penlight directly in my eyes. "Does that help you spot it?"

I don't reply, staying still and waiting for him to point the light elsewhere. I know he will, because he craves attention and wants my interaction. He soon shifts the light to the floor, and I give him a bone to keep him engaged. "Aaron wouldn't be duped so easily into being an errand boy for a complete stranger. Tell me what you've really done?"

"Ah, Tatiana, it's so fortunate for me that you and Joe have such a penchant for offending and infuriating people. It's proved very helpful."

You won't be feeling so fortunate when you get caught. Huh, he's wearing different clothes, camouflage pants, sturdy moss-colored boots and a heavy tan windbreaker.

"It was child's play for me to dupe him," he boasts. "I bought Aaron a few drinks at the local pub, then started spinning a tale about what a huge pain in the ass Joe's been in my life. He readily commiserated with me and soon became willing, even thrilled, to help. By now, he'll have handed my letter off to your friend Hannah, who in turn will pass the note on to Ms. Mizuikova. I see a boatload of money in my future. Freedom, here I come!" He pushes up the cuff on his jacket and glances at his watch.

"I wouldn't be celebrating yet. A certain serpent will rain down terror on your freedom." *Wow!* Maybe my father did know what he was doing when he hired her, someone who is distrustful of everything and leaves no stone unturned. "Your money-boat isn't seaworthy. The moment Ms. Mizuikova receives that letter, she'll poke holes in it by thoroughly questioning everyone involved, leaving your dinghy upended."

"Emotions have a strange way of twisting people's thinking, Tatiana. Ms. Mizuikova's hatred of Joe and you will warp her judgment. She'll have little desire to rigorously investigate a love letter that vindicates her scathing opinion of him, and by the time she realizes her mistake, it'll be too late," he smugly recites his own version of events.

Doubt seeps in. Ms. Mizuikova does hate Joe ... and me. It just might go down that way.

My captor shines the light at his wriggling fingers onto the wall, creating demented shadows that flicker and come to life like an old-time flip book. One such image looks like a roaring tiger with its mouth gaping open. His spooky shadow show ends just before the tiger reaches me, when he focuses the light back in my face.

"Where would you have gone, Tatiana, if Joe had bought you that car? Or should I call you Anna Alkaeva, from the Russian verb 'alkat,' meaning to wish? But Joe won't do anything ever again for you, will he?"

His provoking statements and hideous invasion of my feelings for Joe finally break through my defenses. "Shut up!" I fire at him. *Calm down, Tatiana, and stay focused. That's what's going to keep you alive.*

"Have you read about the effects of light deprivation, Tatiana? It alters the brain systems that regulate mood, making you dead to everything but your torturing anxieties and horrible despair. I wouldn't recommend it. Answer me this one question, and I'll place the penlight on top of those boxes for you. What do you wish for?"

"I wish a meteor would plunge from the sky and burst you into a billion inconsequential pieces of nothingness." Gathering up what spittle I can, I launch it at him. Direct hit! The glob stickily drips down his cheek.

Reacting like a madman, he seizes the plastic water bottle he brought, furiously spills it over the top of my head, then savagely throws it across the room. "You should have wished for water, Tatiana." He wipes my spit onto his sleeve then storms away, so upset he forgets to stick the gag back in my mouth.

The stupid fool, he actually thought he could break me down. Well, I won that round. *Gavno!* He's turning back. Stomping over to my corner, he ties the gag in place wickedly tight and heckles me one last time. "Since you love music so much, Tatiana, I'll leave you with this little ditty to contemplate during your long, dark, solitary confinement. *'Bolt the lock, put out the light, he's not coming, for you tonight. Fight the fright, place out of sight, he's not coming, for you any night. Filled with spite, nothing seems right, he is not, your shining knight. Ta, da, da, da, da, it's getting darker, ta, da, da, da, da, it's getting colder.'* " After childishly jeering a few more *'Ta, da, da, da, das'* at me, he struts away with a bounce in his step.

The water he poured over me is dripping down from my hair onto my face, and I lick up as much as I can. *You were careless, Tatiana. You shouldn't have provoked him until after he gave you water.*

The sound of the lock clicking shut presages my extended internment. Darkness ... chilling darkness ... more unsettling darkness. *Ta, da, da, da, da, it's getting darker. Ta, da, da, da, da, it's getting colder.* His juvenile gibes creep under my skin. No, I'm not going to let that happen. Viktor, or whatever your name is, you will not be the victor of me, reason being I can hold my own against brutal odds. I've lived most of my life in solitary conditions with bitter winds whipping the barren tundra of my soul. I am a survivor.

As loud as I can with the gag in my mouth, I chant my own lyrics. *"Here I fight, the coldness of your spite. Joe will dare, the bitter night, with all his might, for he is, my shining knight."* The muffled words are almost immediately swallowed up in the gloom. Darkness ... chilling darkness ... more unsettling darkness. Minutes pass, slowly blending time together, until everything is a conglomerate mass where I can no longer perceive one moment from the next.

"Joe, help!" I scream, beseeching him with all my heart to rescue me. Frantically I struggle to free my hands and legs, but my frenzied actions only cause me to bang my head against the wall, sending out jolting bolts of light. I blink rapidly. Rows and rows of books appear with one placed on the table before me, displaying a haunting photo of a deserted schoolyard. The grief of hundreds, yet the grief of one. *Zzzap!* I'm sitting on the edge of my mama's bed, staring down at the image of the destroyed Chernobyl Nuclear Power Plant. The grief of hundreds, yet the grief of one.

I can't withstand the breadth and depth of this despair. Panic-stricken and desperate, I start crawling far inside myself, shutting everything else out. But before I totally disconnect, I make one last plea. "Sing ... Joe ... Sing very strong!"

Chapter 37

Our bodies are not eternal
But the universe is
Each insight we gain guides us
Empowers us
And we learn to *Sing its Song*
To understand
The ethereal experience of being

Joe
Saturday night, October 13 to Sunday, October 14

I should have helped Anna. I don't know if I can survive her going missing too.

Sing.

Where's that coming from? I look around. Lance has his headphones on.

Sing very strong.

That was so faint it was almost indiscernible, yet I recognize it as a line from a poem Kluge wrote down and sent to Lina. Life just keeps getting weirder by the minute. Jeez, I'm feeling panicky. Now nothing, nothing at all. *Whoosh.* My emotions rush back in, extreme worry and fear for Anna. I'm majorly screwed up.

My head and hurt hand are throbbing. Popping some of the pain pills I got from the hospital pharmacy, I lie down and attempt to rest. *Where are you, Anna, and how will I be able to find you?* For over three years I've been trying to track down Lina, and she had no money or resources to help her disappear. I don't stand a chance of finding you.

Exactly, Joe. Anna has money and resources, so does her father. Let him hunt her down, you stay out of it.

The pills start taking effect and I drift off to sleep.

I'm in the Hibbing High School auditorium. Mozart's E Minor, Albert Einstein's favorite violin sonata, is being performed. Halfway down the aisle, standing and beckoning to me, is the man himself, the one with the shock of untidy white hair and a dead pipe clamped between his teeth.

"How does music help you think, Mr. Einstein?" I ask him as I approach. "Because all it does for me is cause torment and guilt."

"Life without music is inconceivable, Joseph." He places a firm hand on my shoulder. "I live my daydreams in music, I see my life in terms of music; it's the lightning rod that helps me understand our complex existence."

"How? How does music do that for you?"

"Come, sit and listen to Mozart with me." Mr. Einstein leads me into row J and settles into seat 47, the seat that's always down, the haunted one. Giving me a reassuring wink, he pats the spot next to him. As soon as I'm settled, he excitedly whispers in my ear, "Mozart's music is so pure that it seems to have been ever-present, just waiting to be discovered. This sonata will help you understand the universal harmony of the universe, Joseph. Listen and allow it to flow through you."

I sit and I listen and I try to let the music flow through me, but all that does is get me more frustrated. "Music isn't helping, Mr. Einstein. It's only complicating things."

He turns his head to look at me. "We will never fully comprehend our existence until a unified theory of everything is discovered."

Physics, good, this is a far more manageable subject. "We're closer to that now than ever before, Mr. Einstein. It's called Superstring theory."

A sparkle twinkles in his eyes. "Tell me about this theory."

"It construes the world as inconceivably tiny strings whose different vibrations underlie everything that happens. All particles are thought of as one-dimensional ribbons of energy on which the fundamental forces of nature hum."

"Hmm, strings?" Folding his hands together, he hums to the music and rhythmically taps his fingertips in time to Mozart.

"Yes, strings," I say. "Just as violin strings resonate different frequencies to produce musical notes, these energy strings resonate different oscillatory patterns to produce the preferred mass and force charges."

A gleeful smile lights up his face. "So by replacing the point-particle material constituents with strings, the incompatibility between quantum mechanics—the laws of the small—and general relativity—the laws of the large—is resolved. Very interesting. My theory of relativity was a musical thought, also." He ponders this more deeply as the solo violinist's strings resonate over the orchestra's accompaniment. "This new theory, this Superstring theory, is it a world of many dimensions?"

"Yes, and parallel universes."

He hums a few more measures. "Strings. Why didn't I think of that? It seems so obvious now. Strings are the universal instruments that perform the grand, cosmic symphony which embraces the realities of our experiences! Did you know I was six when I learned to play the violin, Joseph? My Lina is my constant companion."

"Lina!" I exclaim.

"Yes. My violin is named Lina, and she travels the globe with me, shaping the way I view the world." He releases a bittersweet sigh. "Most of the joy in my life comes from her."

"I have a Lina in my life too, and the meaning of her name in Russian is illumination. Why do these bizarre coincidences keep happening to me,

Mr. Einstein?" There must be an explanation, and you're the one person who can decipher it.

"God does not play dice, Joseph. Nothing is left to chance," he solemnly replies.

"God shouldn't be playing with dice," I petulantly state.

"God's not the one rolling your dice, Joseph, Lina is."

"Lina!" I bolt upright in bed and look over at my alarm clock. One fifteen a.m. *Crap!* Another night of little sleep, since I'm not staying here to chance having more haunted dreams. As quickly and quietly as possible with one hand immobilized, I try to change into running clothes. That's not happening, so I settle for moderately fast and noisy.

"What are you doing?" Lance mutters, as he squints at his clock. "It's one—"

"I know how to tell time, Lance. We've had this discussion before." Three days ago, in fact, at three thirty a.m., Thursday, October eleventh, and I remember exactly what was said. *Where are you going? It's three thirty in the morning.* Do you remember that? Of course not, only I'm cursed with this freakish ability.

I rush outside, where the cold air invigorates me, and start jogging. Almost immediately, my hand begins to send out wicked pain messages with a vengeance. I actually don't mind, that's better than another shitty dream, at least this is a demon I have a little control over.

The night sky is bright enough to see by, so I glance back numerous times to check for any sign of the serpent's patrol. No one. Entering the arboretum, I turn to the right and run past Postage Stamp Prairie then head along the southernmost part of the route. Reaching the turn north that borders Spring Creek Road, I slow down and walk for a bit. The physical exertion, combined with the solitude, is calming, even with my hand complaining about the rough ride.

I gaze upward and start mentally checking off the constellations I can spot. Pegasus is easily recognizable right near Andromeda. Fomalhaut, the brightest star in the constellation Piscis Austrinus, is close above the southern horizon. Since it's in a part of the sky that appears largely empty, Fomalhaut is often called the Lonely One.

The Lonely One. Lina. Anna. Me. We're all lonely ones. Tonight though Fomalhaut isn't lonely, lingering nearby is a companion, Mars, the god of war and bringer of bloodshed. To the casual star watcher, nothing would seem amiss with Fomalhaut and Mars, but to the trained eye, Mars appears especially red this evening, blood red. The hairs on the back of my neck stand up, leading me to intently scan my surroundings. At the edge of the nearby the tree line, I notice some movement, two men darting over to the trail.

"Take it easy, boy, and no one has to get hurt," the taller man warns me in Russian as they approach. I recognize his voice, one of Medusa's brood.

With my broken hand, I can't fight off both of them at once, but I bet I can outrun them. Doing an instant about-face, I take off at full speed. *Damn.* There's another one of them blocking the path going this way. No choice now. Bolting into the fellow, I hit him full force in the chest. Unfortunately, he manages to grab hold of my sweatshirt, and we both tumble to the ground. Cupping and rolling out of his grip, I'm able to protect my hurt hand.

Before I can regain my footing, the other two are on me. They start shoving me north along the path, and just before the split, we turn to the east and come out on a road. A dark sedan is waiting. After pushing me into the back, one bodyguard gets in on each side of me and the third jumps into the front passenger seat. The driver swiftly swerves onto the road and guns it. We ride a few miles without passing a single car then turn into a hidden drive at the apex of a sharp curve and pull up to a small, janky house that's set way back from the street. The second we're parked, the front door of the house opens and Medusa slithers out, coiled and ready to strike.

"Not so clever now are you, dvoyechnik?" she hisses at me as I'm dragged up the front steps, hustled into a small room that's empty except for a few wooden chairs and forced to sit down. With a delighted smirk, Medusa circles me. "Your ineffective ploy of manipulating Aaron into being your messenger boy didn't deceive me for a single moment. Both Aaron and Hannah spilled the beans a mere five minutes into my questioning. After that, I quickly ferreted out the ransom message hidden within that sickly love letter."

"Quit spitting Parseltongue at me, Medusa." What the hell is she talking about? And what the hell does she think she's doing ambushing me and holding me hostage here?

Her eyes gleam wickedly as she continues to circle round me. "You never fooled me. Right from the first time I set my sights on you, I saw through your charade. It's all over now. Spill your guts and tell me where Anna is."

I'm not telling her a damn thing. She wouldn't know the truth if it slammed her in the face. And after hearing the tales she told Hannah and Lance about me, I'm sure she'd just twist the truth into whatever form fits her current needs best.

"Tell me where Anna is," she yells again. Her shirt is bunched up around her waist and her clothes are wrinkled. She's probably been trying to sleep in them. Her hair has numerous strands out of place and there are dark circles under her eyes. She looks like shit. And when people feel like that, they make mistakes. I know I sure have.

"You really screwed the pooch this time, Medusa. Not only did you lose track of Anna, you kidnapped me. In this country, you can't abduct people and get away with it. In case you didn't know, it's illegal." I doubt this will sway her at all, but it's worth a try.

The glint in her eyes narrows marginally. "Tie him up and stand watch. I'm going to get some rest until Mr. Alkaev arrives. Wake me up at six a.m. unless his flight is delayed."

Mr. Alkaev? That must be Anna's father. So, is his real name Mr. Alkaev or Mr. Alliluyev? Or are both of them aliases?

"Where is he?" A loud, demanding voice wakes me from a fitful doze.

"He's tied up in the back room, Mr. Alkaev," Medusa respectfully replies.

"What else have you found out since we talked last?"

"We retrieved Anna's phone from a trash bin near a bus stop on Carleton campus and found three sets of fingerprints. Anna's, Carleton's and Arkady Petrovich Markow's. Markow grew up in Moscow and was conscripted into the military when he was eighteen. After serving a year with the ground forces in Chechnya, he got out and later joined the Solntsevskaya Bratva, a gang that operates drug smuggling, arms trading, prostitution and kidnapping rackets."

Keep talking in that loud voice that lets me hear you through the walls, Medusa. I appreciate the update. Was this Arkady Markow the one who wrote those letters to Anna? If not, why are his fingerprints on her phone? And if he did kidnap her, why would he have left that phone for them to find?

"The Solntsevskaya Bratva doesn't operate in the United States," Mr. Alkaev tersely states. "What else have you found out about Carleton?"

"Not much. He has an arrest record, but we already knew that. We discovered Anna's purse in the same trash bin as the phone, and inside her purse there was a small listening device. We questioned a number of drivers from the routes that serviced the bus stop near that trash can, one made mention of a man and a woman who seemed to be talking in Russian and boarded his bus, but he couldn't confirm the woman was Anna. The lady he saw looked different than the photo we showed him."

A man and a woman? Is this part of a plan Anna concocted, or was she really abducted? *Keep spilling your guts, Medusa.*

"Where did they get off?" Mr. Alkaev demands.

"At the downtown Minneapolis stop, South 8th Street, near the IDS Center. The driver said the woman seemed to be having a little trouble walking."

"Where did they go after that?"

"We're still working on it," Medusa hesitantly answers.

"Show me the ransom note."

"Yes, sir. The note has four sets of fingerprints. Carleton's and Markow's we were able to identify. The two others we believe to be Anna's roommate's, since she handed me the note, and Aaron Edward Hull's, the Carleton student who gave the note to her. Hull was born in O'Fallon, Illinois, has no criminal record, doesn't speak Russian and has no connection to Markow. It appears he had no idea it was a ransom note."

Better hope I don't see you before you see me, Aaron. That prick will do anything to get at me.

"How did Hull get that letter?" Mr. Alkaev grills.

"He claimed a girl named Mary Anne gave it to him. He said she was an ex-girlfriend of Carleton's who wanted to break up his relationship with Anna."

"What's the full name of the girl?" Mr. Alkaev's getting frustrated.

"He didn't know. We're still looking into that."

"Did you get a detailed description of her?"

"It was rather nondescript," Medusa admits, a nervous edge in her voice.

"Doesn't that seem strange to you?" There's a chilling reprimand wrapped in his question. The serpent has no answer. "Your only task was to keep my daughter safe, and you've failed completely, Ms.

Mizuikova."

Heavy footsteps approach, the doorknob turns and a tall, imposing man enters. His hair has a wave to it, and like Anna's, it's deep black. Their features are similar, but his are sharper and more prominent than Anna's. He slams the door to keep everyone else out, crosses the room in four long strides and towers menacingly in front of me, radiating intense hatred.

"Where's my daughter?" He begins my interrogation through gritted teeth. When I don't reply, he strikes me so hard he almost knocks me over. "I'm not to be toyed with, boy. Where is my daughter?"

I remain silent with a defiant look on my face as he strikes another blow that busts my lip open. I spit the blood at him, but it falls well short. He takes to pacing the room. After a few circuits, he pulls over a chair and sits directly in front of me. "I will find her, even without your help, so I strongly suggest you cooperate. Tell me where you're keeping her, or things will get far worse for you."

I spit more blood at him and manage to dirty the collar of his shirt. "You won't find Anna unless she wants you to. She took off to get the hell away from you!" It feels good to shove that truth down his throat, and it feels even better to see the pained grimace that flashes across his face. From what I know so far, Anna's version of events, her father treats her like she's property.

Surprisingly, he doesn't hit me again. The anger in his face is no longer totally obscuring the fear beneath it. "If Tatiana had run away, she wouldn't be so sloppy as to throw her purse and phone in a trash can where it would be discovered so quickly. Nor would she involve other people. That would only complicate everything for her and increase my odds of tracking her down. She is too cold, calculating and straightforward to risk that. Or at least she was."

He's not saying this to explain anything to me, only talking it through to himself. But, he's right, Anna wouldn't do that. She did get abducted!

"Your fingerprints are on the ransom note and her cell phone, Carleton. You get one more chance. If you tell me where my daughter is, and she's unhurt, you'll live to never see any of us again." He "graciously" throws me a lifeline, one that can be easily taken away.

My gut reaction is to tell him nothing. However, if Anna has been taken, which seems highly likely, not cooperating with him will only waste precious time that could be better spent searching for the real kidnapper. "I don't know where she is," I admit. "I didn't write that note and I'd never do something like this to Anna."

The door opens and my nemesis serpent slithers in. "After further questioned, Hull admitted that a man matching Markow's description gave him the ransom letter. We also got access to Carleton's computer and used a hardware keylogger to open his emails. He received one yesterday from a Viktor Paskhin. It read, 'Joe, we got the package.' Viktor Paskhin served in the same ground force unit in Chechnya as Markow, but he was killed over three years ago. We're in the process of trying to find out where this email was sent from."

Anna's father furiously rounds on me. "You lying son of a bitch, tell me where my daughter is."

How do I always get caught up in the middle of this shit? Well, he's not going to believe anything I say now, so the wiser move is to keep my mouth shut.

Medusa hisses annoyingly to regain Mr. Alkaev's attention. "It's possible that Hull—"

"Stop following that dead lead and find out where Markow took Anna after they got off the bus."

One of the underlings strolls in. "Excuse me, sir. We just finished questioning a cab driver who remembered driving a woman and a man with a Russian accent. He picked them up outside the IDS building in Minneapolis and dropped them off at Currie Ave. There's not much there, just an exotic-dance club, a pawnshop, a church and a six-story homeless facility. Right down the block, though, is a bus station. We'll start questioning bus drivers from there."

"Did the cab driver see which direction they headed after he dropped them off?" Mr. Alkaev asks.

"No, but he confirmed the woman was having a great deal of trouble walking, so they couldn't have gotten far unless they got another ride."

"Untie Carleton and bring him along. We're going to Currie Ave."

Chapter 38

Once in infinite existence, immeasurable in time and space,
a spiritual creature was given on his coming to earth
the power of saying, "I am and I love."
Once, only once, there was given him a moment of love.
What is hell?
I maintain it is the suffering of being unable to love.
Fyodor Dostoevsky

Tatiana
Sunday morning, October 14

My captor enters the room, turns on his flashlight and walks over to me. "It's so dark in here, Tatiana. The only way to find you is by the stench of your urine. How can you stand it?" Leaning over, he takes the gag out of my mouth and waits. "Not answering questions today?"

Not for you.

"Shame, I brought you some food. You've got to be hungry by now." Setting the flashlight on the concrete, he takes a bottle of water out of the white plastic grocery bag that's looped over his wrist and pours some of it on my face, hoping for a reaction. I don't move a muscle. "Well, guess I'll eat it for you, then." He pulls an egg salad sandwich out of the bag, unwraps the plastic, takes a huge bite and chews loudly.

He quickly tires of eating. "On my walk over here, I saw something captivating carved in the sidewalk. 'I am a sojourner in your universe of solitude. Wandering the caverns of your soul, I feel nothing, yet I feel everything.' That mirrors our situation, don't you think?" Using the tip of his boot, he nudges me with his foot. "What's wrong with you?" He gives me a kick in the ribs and I emit a low groan.

"Ура! She's not mute," he excitedly exclaims as he spits a bit of eggshell onto the floor. Leaning over, he picks up the bottle and pours some water in my mouth. I eagerly gulp down a few swallows and my body betrays me, letting out an appreciative sigh.

"Human emotions, they're so hurtful. It drove poor Viktor mad, you know, that you were never able to love him. T'fu, enough of the past, we must live in the present. Yet how I wish I could have seen your grandfather's face when he found out that *his* money will be paying my way out of the Solntsevskaya Bratva, the very gang that murdered his favorite billionaire comrade, the one who was hell-bent on exposing Putin's links to the Mafia. But alas, instead he'll be rolling in his grave with me dancing on top of it." Gleefully whistling a popular Russian dance tune, he starts giddily prancing about.

My grandpapa's grave? Is Grandpapa dead? How long have I been stuck in here? "What day is it?" My voice is so raspy that *I* can barely make out what I said.

"That's not important, Tatiana. You must live for the moment. Tsk-tsk, you should see yourself, you look so wretched and pathetically weak. It's on account of you succumbing to that treacherous emotion, love."

Love? Joe! He's nearby, I can feel him. "Jooooooooooe!" I yell as loud as I can. "Heeeeeeeeelp!" I screech in as high a pitch as my vocal cords will allow.

My captor snatches the gag off the floor and brutally shoves it back in my mouth, but he's too late. I've already broadcast my location, and Joe's coming for me.

Chapter 39

Joe, age seventeen

"Joe, I got a really strange letter in the mail, postmarked from here in town. It didn't have any return address and wasn't signed. It's kinda weirding me out."

"Do you have it with you?"

Lina unfolds a piece of paper and hands it to me.

Dear Lina,

I need you to know. You are the spark of life that keeps me going. Someday, you will understand. Until then, I leave you this.

Sing for me and I'll be there,
if you sing me strong.
Sing when all the day is dark,
and every star has gone.
Sing when all the air is cold,
and home's a far-off spot.
Sing for me when others won't,
'cause they have long forgot.

Only sing very strong.
And shine, shine, shine with the song.

Others will never know,
how you reached out to me.
You crossed the cosmos,
and stretched eternity.
Only you and I will know,
this spooky entanglement of ours.
That holds our souls together,
through space and time and stars.

Only the two of us
will know this secret.
We are immortal.

I immediately recognize Kluge's handwriting. Damn him! Later in the day, when I get the chance to unload on him, he's going to experience some choice German swear words he's probably never heard before. It practically kills me to wait until after school to confront him, but if I skip any more classes, my dad would probably kill me. At long last the bell rings and I race out of the building, heading straight over to Kluge's place.

"What the hell do you think you're doing?" I shout the second I'm inside, my anger reaching every crevice in this lousy hovel. As I storm into his den, he raises his head, but doesn't turn to look at me. His back is ramrod straight and his right hand is posed in midair, gripping his favorite fountain pen.

"Ellie showed you the letter." His voice is mixed with sadness, bitterness and regret.

Moving around to the front of his desk, I furiously swipe away the papers laying there. "You lying two-faced, sadistic bastard. You brainwash me into never revealing anything about you to Lina, then you pull this stunt."

Kluge continues staring straight ahead, almost as if he can see through me. "I'm dying, Joseph. I have terminal cancer, and I want her to have something of her true heritage before there's nothing left."

"And this half-assed idea is your solution? Well she didn't get any sense of family ties from it, only a twisted feeling of being stalked." I glare especially hard at him. I've been glaring ever since Lina showed me that letter. My face is going to get frozen this way.

He slowly tilts his head to connect with me, eye to eye. "Her grandmother—my wife, Syntyche—wrote those words to me a few days before she died. I … I needed to pass them on to Ellie." Tears form and a few run down his cheek. The only other time I've seen him show such tender emotion was many years ago when I was repeating his own Russian words back at him, and that was only for a few seconds.

"Promise me, Joseph, promise me you will keep Ellie safe."

Joe
Sunday morning, October 14

After we pull into the lowest level of the parking garage on Currie Ave, everyone wraps their scarves around their faces and pulls their hoods over their heads before getting out of the car. Everyone except me, that is. My face is left totally exposed for the entire world to see. Someone needs to be left to take the blame. Anna's father divides us into three groups of three each, not counting me, and assigns each unit a different building along Currie Ave. I'm grouped with her father and two of the newer security personnel who arrived with him.

We're the first unit to exit the garage. The wind is whipping up dust into the air, and while it's not close to the record low of minus sixteen for this date, it's still pretty cold. Although it's miserable for me, it's very fortuitous for Anna's notorious search party; no one will think twice about their heads and faces being completely covered up.

My muscles tense as Anna's emotions bump at me. She's somewhere nearby! My first instinct is to dart off and find her, but that's not possible or practical with my immobilized hand, no weapon and a formidable squad surrounding me. We're all waiting, very law abidingly, for the pedestrian "go" signal. When it turns green, we briskly amble across the street and along the sidewalk bordering the Salvation Army homeless shelter. This was one of the many places I rampaged through this summer searching for Lina. As we pass the main entrance, someone calls out, "Hey! Hey!"

I spot a guy about my age smoking a cigarette and waving his hand to get my attention. His straight, jaggedly cut, black hair oddly frames his round face. Straightaway, I remember him. He was one of the people I showed Lina's picture to.

"Remember that girl you were looking for? I saw her here a week ago," he yells.

Before I can respond, Anna's father grabs my injured hand and menacingly whispers in my ear, "Keep moving and keep quiet."

No way am I keeping quiet and passing up this golden opportunity. I've been chasing down rabbit holes trying to find Lina for three years without any success. "Where did she go?" I shout.

"Don't know," he volleys back. "She came in for a meal, then left."

"Not another word." Anna's father twists my hand, sending pain shooting up my arm. We reach the end of the building, round the corner, then halt by a side door. One of Mr. Alkaev's henchmen expertly picks the lock while we form a close semicircle around him. In short order, we're inside.

"You two start searching this floor. I'll head up to the next level." Mr. Alkaev prods me along the hallway and into the stairwell at the end. Acute anxiety, which I'm sure is linked to Anna, hits me hard in the gut. Is it coming from above or below? I can't pinpoint it. Halfway up the flight of stairs, I swear I hear Anna call my name. A millisecond later, I double over like I was sucker-punched.

"What's wrong with you? Get up." Grabbing my arm, Mr. Alkaev starts hauling me up the stairs.

"Didn't you hear that?" I gasp. As he continues forcing me upward, I intuitively feel we're going the wrong way. "Downstairs, we need to go down."

"Keep quiet," Mr. Alkaev rebukes me.

"Heeeeeeeelp!"

"That was Tatiana!" her father exclaims. Totally disregarding everything, including me, he heedlessly bolts down the staircase. Racing after him, I catch up on the last step before the basement landing where he's standing motionless, analyzing the best course of action. To the left is a metal door that's slightly ajar and to the right is a large open area with a furnace and a maze of storage units. Talking into his collar, which has a small communication device attached, Mr. Alkaev quietly redirects the two men above us. "Dmitry, get to the basement to cover us. I heard Tatiana scream and it came from here. Anton, notify the others where we are, then make your way down."

Focusing left, then right, I try to emotion-locate Anna. *Which way are you?* Before I have a chance to tune in and pinpoint her position, Dmitry's on the step above us with his gun in hand.

"Cover me. I'm going through the door," Mr. Alkaev says. "If Markow's in there and you can get a clean shot, take him down, but don't kill him. I want him alive." From a holster concealed under his coat, Mr. Alkaev pulls out his gun. They're both holding Russian PSS silent pistols, which are almost completely quiet when fired. On the downside, the magazine contains only six rounds, the ammunition is of relatively low power and it's not effective beyond eighty feet. I know all this because Kluge kept one in his bedside drawer.

Mr. Alkaev kicks the door wide open, shoves me in front of him and places his pistol against the back of my head. It's pitch-black except for the shaft of light that's behind us. *Shit!* We're going in blind, and if Markow's here, he's seeing everything.

"Another step and I kill her," a steely voice warns from a back corner.

We swiftly shuffle sideways. "You've got no way out of here, Markow. If you kill her, it's your death sentence too," Mr. Alkaev coldly threatens back.

"I have a syringe against her neck with undiluted M99. Your daughter will be dead within minutes if I inject her. Therefore, I'm the one who will be dictating the terms here." Abruptly, a bare bulb hanging from the ceiling illuminates the room. Squinting against the sudden brightness, I'm able to make out Markow hunched over Anna's slumped body.

"The others are on their way. They'll be here shortly," Anton announces as he trains his gun on Markow.

"It's over. Step away from her," Mr. Alkaev commands.

"You're still not in a position to tell me what to do, Mr. Novikov," Markow confidently touts.

What are they waiting for? Someone must have a clean shot. Markow needs to be taken down now. The barely audible clicks of the gun mechanisms sound. Two shots are fired.

"Tatiana!" Anna's father darts forward to her. Medusa and her team race into the room, jolting me out of my trance-like state. The serpent heads straight for Anna and I trail behind. She bends down to check her pulse. "Anna's still alive," Medusa states, all businesslike, then starts removing the bindings from her hands and feet, methodically checking for wounds. Finding nothing requiring immediate attention, she takes an IV out of the backpack she brought and inserts it in Anna's arm. All this while, Mr. Alkaev has been locked in shock, and I'm not doing much better. We're both helpless spectators.

"Ahem," Dmitry clears his throat. "Markow's dead," he hesitantly says, knowing he screwed up. "He didn't inject your daughter, though. The needle's still full."

With Markow out of the picture, Anna's father lurches at me to vent his rage. Slamming me against the wall, he clutches my throat. "What have you done to my daughter?" he howls, sounding like a half-wounded, half-rabid wolf.

"I wasn't part of this," I choke out.

Seizing my injured hand, he pummels it into the wall with a vindictive fury. The jarring pain incapacitates me. He's all rabid wolf now. An instant later, a different kind of agony rips through me.

"AHHHHHHH!" Anna's torturous cry pierces the room, triggering her father to instantaneously sprint over and cradle her close. He tries desperately to console her, but nothing he does quells Anna's tormented screaming. If anything, he's making it worse.

Half demented myself, I scramble over to her other side and grasp her hand. The instant we touch, she becomes deathly still, yet her internal agony won't quit. It's spiraling out of control.

"We need to leave," Ms. Mizuikova flatly states.

Mr. Alkaev continues feverishly begging, caressing, squeezing and gently shaking his daughter. None of it has any effect. He morosely leans back and stares at her motionless body, then it finally registers with him that I'm holding Anna's hand.

"I'm going to kill you for what you've done to her!" he booms in a thunderous voice that's pure malice. He's at his wit's end. Having found his daughter, he's now powerless to do anything for her as she slips away from him to a place where he can't follow.

"We risk being arrested if we stay here any longer," Ms. Mizuikova loudly insists. "We need to go."

Despair seeps through every cell in Mr. Alkaev's body, changing his entire demeanor. His shoulders slump and his head droops downward.

Seizing the opportunity, I squeeze Anna's hand in mine and kiss her passionately.

Chapter 40

Tatiana, age ten

"Why do you have to go? I don't understand, Grandpapa." I'm tugging on his coat and staring up past his grizzly, gray beard.

"Someday you will, Tatiana. Until then, sing for me and I'll be with you, only sing very strong." Reaching out, he touches my forehead and gently brushes my hair out of my eyes.

"How do I sing very strong, Grandpapa?" I hold the bottom of his heavy coat tight, not wanting him to go.

"By singing even when all others have forgot," he answers in his gravelly voice.

"I don't know if I can sing that long. Why can't you stay here, Grandpapa?" No one tells me anything. It's not fair.

"It's not safe for me, Tatiana." Bending down, he pulls me into his arms and hugs me so close that it's a little hard to breathe. When he finally lets go, Grandpapa's eyes are watering. I'm so confused. I don't understand any of this.

Pulling a piece of folded paper out of his pocket, Grandpapa hands it to me. "A long time ago, someone very dear to me had to go away. She wrote down these words and gave them to me. Keep them close, Tatiana, and they will help you too," he whispers.

I unfold the paper and start reading.

Sing for me and I'll be there,
if you sing me strong.
Sing when all the day is dark,
and every star has gone.
Sing when all the air is cold,
and home's a far-off spot.
Sing for me when others won't,
'cause they have long forgot.

Only sing very strong.
And shine, shine, shine with the song.

Others will never know,
how you reached out to me.
You crossed the cosmos,
and stretched eternity.
Only you and I will know,
this spooky entanglement of ours.
That holds our souls together,
through space and time and stars.

Only the two of us
will know this secret.

That doesn't help at all. It won't stop Grandpapa from leaving.

He tilts my chin upward to regain my attention. "I have something else to give you." Removing the letter from my hand, he carefully refolds it and tucks it into my sweater pocket. Hunching over very close, he murmurs softly in my ear. "No one can ever know of this. Do you understand?" Leaning back slightly, he stares intently into my eyes.

I nod yes. This I can understand. I'm very good at keeping things to myself.

Grandpapa places a small cloth satchel bound by a black string into my left hand. Untying the string for me, he lets me peek inside. Three small stones sparkle and twinkle at me. "These are called Silver Bears, and they will be able to help you when I'm not here to. Just like your pet bear, Tatty, they are very special and very powerful."

The satchel feels heavier in my hand now, even though it's light.

"Listen carefully. These Silver Bears are very valuable and should only be used when you have no other choice. On the back of the poem I just gave you is a phone number for my most-trusted associate. Call him if you're ever in danger."

"Tatiana!" My father's voice booms through the house.

"Quickly, hide them where no one but you can ever find them." Grandpapa urgently shoos me out the back door. I hear him arguing with my father as I scurry away and crawl between the large bushes growing against the house. After a few moments of really hard thinking, I know exactly where to put my Silver Bears.

I wait until my father stops calling for me and drives to work, then I sneak into the cellar and tiptoe over to the locked closet in the back corner that's hidden by a tall pile of boxes. Father tried to hide it, but I know the closet is there and I know where the key is, tucked inside my mama's old tackle box, taped to the bottom of a container of bobbers. Using all my weight, I shove the boxes aside, then retrieve the key. As silently as possible, I open the door and squeeze my way to the back while Mama's favorite coats and dresses rustle around me. Closing my eyes, I breathe in her scent and crouch in the darkness until my muscles get stiff.

"Tat, where are you?" I hear my oldest brother's footsteps above me as he searches the kitchen, opening the cabinets I sometimes hide in. *Blin.* Why does he have to be looking for me now? Most of the time, he ignores me.

I quickly open a small shoebox that contains a fancy pair of my mama's shoes. Before hiding the Silver Bears away, I untie the string for one last look. The diamonds still twinkle at me, even in this dark closet, and a weird shiver shudders down my back, even though I'm not cold. I retie the string and tuck the satchel, along with the poem, inside a small wooden puzzle box my mama taught me how to open. She was going to give it to me on my seventh birthday, so I feel like it belongs to me. The puzzle box fits snugly inside the shoebox which I stuff inside another larger, cardboard carton and cover it with scarves. After relocking the door, I shove the boxes back in front. "Stay here, Tatty, and keep these other bears safe," I whisper to her before putting the key away and heading upstairs.

Tatiana
Sunday morning, October 14

Tingling, a distant buzzing, a vibrating energy on my lips. Warmth, a quiver of heat, a swirling, powerful, emotional wind blustering through me, fanning the spark. The flame exponentially expands to the point where it becomes a roaring blaze blasting through my body. Tipping point. The fire breaks down into isolated hot spots, then recedes until it's just smoldering patches. The aftermath. A vast array of embers and ash glowing with a comforting halo of color. A breath softly stirs the embers, causing them to take flight, circling all around me. During their journey upward, they blink in and out, almost like they're winking at me.

Joe! Joe's here in this secret world with me, this world filled with life amid the ashes of destruction. Our souls unite, two as one, clinging together in an intricately entangled embrace while our unexplainable connection streams lifetimes of passion wordlessly between us.

"Tatiana." My father's voice arrogantly intrudes.

Outside feelings start poking their way in. My head is pounding, my arms and legs are so sore they've gone on strike and my pulse is spasmodically beating. But my hand … my hand is locked in Joe's with a soul-saving grip and my lips are joined with his. Prying open my eyes, I see him pull away to gaze down at me. My wondrously, beautiful blue eyes of Siberia.

"I sang out when all the day was dark," I rhythmically chant to Joe. "And I sang strong." *I understand now what it means to sing strong, Grandpapa.*

"Tatiana, look at me. We need to get you out of here."

I'm tired of you always telling me what to do, Father.

"Can you stand?" He sits me up and Joe releases my hand. Father then wraps his coat around me and lifts me into his arms, not bothering to find out if I can walk on my own. "Anton, collect the spent casings and dig the bullets out of Markow. Dmitry, move the boxes over the bloodstains. Maxim and Sergei, bring the cars around, and one of you get a blanket for Tatiana. Leave in small groups and make sure your faces are concealed. Ms. Mizuikova, coordinate with your team and have them go to the drop-off location to see if anyone shows up," my father regimentally barks out orders.

You should have listened to me, Markow, and let me get off of that bus alone.

Everyone jumps into action, except Joe. I stare at him with a comforting smile and newfangled amazement. He found me! *You were right, Grandpapa. The words in that poem did help.* " 'Others will never know how you found me. You listened to the cosmos and stretched eternity,' " I whisper more of the long-forgotten verses as they arise in my memory.

Joe's features furrow with confusion and my comforting smile fades. As the fogginess in my brain disperses a little, I notice his swollen lip, bruised cheek and the blood seeping through the wrappings on his bandaged left arm. What's been happening in his world while I've been trapped down here? Before I can ask, Ms. Mizuikova appears in my field of vision and removes an IV I didn't realize I had, then wraps my head and most of my face with a scarf.

"Do you hurt anywhere, Tatiana?" *What a ridiculous question to ask, Father. My entire body aches.* Taking the exasperated expression on my face as a *yes*, he hurries through the doorway, and with a strength I didn't know he still had, makes his way up the stairs, down a small hall and out a side exit of the building with me cradled in his arms. I peer over my father's shoulder at Joe, who's following like a kicked puppy. What's going on?

A black sedan glides up to the curb and my father places me in the middle of the back seat, sliding in next to me. Joe rushes around to the other side and jumps in a few steps ahead of an irate Ms. Mizuikova.

"Get in the front and don't attract attention, Ms. Mizuikova," my father directs her. She begrudgingly obeys and slinks around the car. The second she's in, we immediately whisk off.

Eww! I reek. Inside the small confines of the vehicle, the stink is horrendous. Ms. Mizuikova leans forward and turns the fan on full blast, then shoves her backpack at my father. He pulls out a water bottle and a package of peanut butter-filled crackers which he passes to me.

I'm wickedly hungry and thirsty. How long has it been since I've eaten? Wolfing down the crackers, I end up getting crumbs all over me and the seat. "What day is it?" I inquire.

"Sunday," my father answers as he unfolds a red and black plaid blanket, draping it around my shoulders and over my legs. It has a musty smell to it, but I'm sure it smells better than I do. After downing the entire bottle of water, I'm somewhat satiated and happily lean against Joe. The warm feeling of his body pressed against mine, combined with the humming of the car, begins lulling me to sleep.

"Tatiana," my father summons me.

"Leave her alone," Joe countermands.

"Tatiana, listen to me. This fellow is not who you think he is. He's been involved in your kidnapping from the start and somehow brainwashed you into thinking otherwise."

Yolki-palki! The high-voltage hatred erupting from both of them jolts me wide awake. "If you really believe that, Father, then why are you letting him sit next to me?"

"He's controlling you somehow, and I don't want you going unconscious again." My father bends forward to yell across me. "Once I figure out what you've done to my daughter, you'll never set eyes on her again."

Wrong tactic to take with Joe, Father. You'll never get him to tell you anything that way. He must have been already trying to strong-arm information out of him; that's why Joe's face is busted up.

"You're the one who will never set eyes on her again," Joe scathingly retorts. "When Anna gets the chance, she's going to beat feet and disappear from your grasp forever."

"Stop it!" I complain. These two words have gotten more use the past few weeks than they have in my lifetime. Except, possibly, for when I was in my terrible two's. And I'm ready to say them again because being the punching bag between their heavy blows, *living* their divisive feelings, is unmanageable, especially after what I've just been through.

My father sits back in his seat, highly agitated and unable to remain silent. "Think back, Tatiana. Do you remember Markow contacting or calling anyone besides Carleton?"

"Joe wasn't part of this," I correct him.

"Markow didn't carry out this kidnapping scheme alone. By abducting you, he was relaying a clear message to your grandfather to stop fighting the Russian regime."

Grandpapa! "Is Grandpapa okay? Is he still alive?"

"Yes, he's fine."

I breathe out a sigh of relief, glad Markov's reference to dancing on his grave wasn't a literal one.

"This ordeal isn't over yet, Tatiana. The Russian regime will not stop until your grandfather is stopped, permanently."

"I need to rest, Father." Beyond exhausted, I collapse against Joe. As I lay my head on his shoulder, the weariness touching every part of my body begins usurping the temporary jolt of 'get-up-and-go' I got from the food and their feud. Yet even though I'm totally spent, a nap seems impossible with all the bad blood flowing between Joe and my father streaming directly through me. Nonetheless, I feign sleep to avoid more confrontations.

The car starts decelerating, presumably to exit the highway, when the driver suddenly swerves to the right. *Beeeeep!* "Stupid Americans," he mumbles as he lays on the horn. I open my eyes and see familiar landmarks go by. After proceeding along a few kilometers, we arrive at and then pass the street leading to Carleton campus. Where are we going?

The driver zips us through a narrow stretch of road in a wooded area before slowing down at the tip of a sharp curve to veer onto a windy dirt driveway. We pull to a stop before a small, weary old house nestled among the trees. It looks even more tired than I feel, and that's saying a lot.

"Stay here," my father instructs me. *Really, Father, where else would I be going right now?* Getting out of the car, he snaps more commands. "Search Markow's body thoroughly before you dispose of it. Get me a report from the unit stationed at the ransom drop and ..."

Click, the doors lock. "Don't take off by yourself, Anna. I'll help you get away from your father if you want," Joe whispers, even though we're alone in the car.

Oh, my spectacularly complicated, blue-eyed Joe. Dreamily, I drift back to that special place of burning embers and twinkling sparks of flame Joe's recent kiss illuminated, that wondrously different reality far from this one. He joins me in the memory, and we fully immerse ourselves amongst the magnificent sense of being and belonging.

A raucous banging rudely shakes our sanctuary. My father is impertinently smacking the roof of the vehicle. "Unlock the doors and get out, Carleton."

Giving my hand a reassuring squeeze, Joe reaches through to the front door and unlocks the car.

"Bring him inside and tie him up," my father immediately orders. Getting in and plopping down next to me, he reinitiates his earlier 'Joe' bashing. "Carleton's fingerprints are all over the ransom note, and just after you went missing he received an email that read, 'Joe, I've got the package.' When I questioned him about this, he was defensive and evasive. He *is* part of your abduction, Tatiana."

My father's tone has softened slightly, a touch of supplication mixed in. Shame it took me getting kidnapped to bring that about. However, that tiny soft spot won't be enough for me to convince him that he's wrong about Joe. *T'fu!* The intense strain from this ordeal has drained my energy; I don't have the strength to fight World War III with my father right now. I just want to rest.

Don't drift off, Tatiana. I pinch myself hard to remain alert. If I fall asleep who knows what in the world my father will do to Joe, or where he'll whisk me away to? I dip into my reserve tanks for a bit of added oomph. "Markow was working alone. He wanted money to buy his way out of the Solntsevskaya Bratva."

"Tatiana, please, think through this with me, reviewing all the *facts*. Carleton has gotten inside your head and twisted your thoughts. He's been using you all along."

He certainly has gotten inside my head, Father, but your assumptions about the mechanics and motives are incorrect. "You're the one twisting my thoughts, just like you did Mama's. She needed you, and you weren't there for her. Her headaches kept getting worse and worse, and you didn't help her." I purposely shove this awkward barrier between us, hoping he'll back off.

He looks away, attempting to hide the pain and bitterness he's feeling. When he turns back, his face is rigid. "I took her to all the best doctors, but they didn't know what was causing her migraines. All these years, I've been agonized over your mother's death, but I can't see anything I could have done differently, and the last thing I need to be dealing with is you parroting your grandfather's accusations."

Chyort! I've stepped into a minefield. Father's deep agony slices into me, causing a vicious headache to erupt. Retching forward, I shove my head between my knees and start rubbing my temples.

"You're getting a headache?" The alarmed panic in his voice sends another jolt through me. "Tatiana, I'm sorry ... I'm sorry I wasn't there for you when you were growing up ... it's just you reminded me so much of your mother, and I couldn't risk it. I couldn't risk you getting the migraines too." Clumsily, he places his hand on my leg.

He's never apologized to me! Half sitting up, I tilt my face toward him. "Did you *feel* Mama's pain when she had the headaches?"

His features crease and he bites his lower lip, fighting the flood of heartbreaking memories as they course through both of us. Without wanting to, he's given me my answer.

"Then you of all people, Father, should understand. When I started getting these gargantuan headaches, the same type Mama got, Joe's been the only one who can help me control the excruciating agony."

A vengeful wrath writhes its way among my father's wrenching emotions, intensifying my already monstrous migraine. Grasping my hair with both hands, I start pulling at it, needing to divert the colossal pain. My father reaches out to embrace me. Unlike Joe's touch, his intensifies the suffering.

Frantically, I pull away from him, struggling to escape all of this. "I need Joe," I hysterically plead as I start to pass out. I'm remotely aware of my father taking me out of the car, then everything goes blank.

IV

Part Four

FOURTH MOVEMENT IN THE COSMIC SYMPHONY OF LIFE

Venus

The goddess of love and beauty

Chapter 41

Joe, age eighteen

I begrudgingly approach the hospital door with a bad feeling in the pit of my stomach. It's been ten days since I visited Kluge, ten days filled with gut-churning heartache, ten days since Lina's dad beat her during one of his drunken stupors, ten days filled with raging fury at her father and no way to act upon it—because I promised Lina I wouldn't go near him—and ten days of wrestling with what, if anything, I should mention to Kluge. If I don't say anything about it, how do I hide it, because it's written all over my face?

I crack open the door. Kluge's body is shrunken into the hospital bed, his skin sallow and drawn. Yet, he hears the door open and glances over. His eyes are shrunk deep into their sockets, but they're still alert. "I've been waiting for you, Joseph," he croaks. "What's happened to Ellie?"

How the hell does he do that so fast? One glimpse at me, and he instantly knows if anything has happened to her, good or bad.

"Her father hit her, didn't he?" Even though I've been grappling with this for days, I have no idea what to say to him. "I've been watching him all my life, Joseph, and I know what kind of man he is. Promise me you will protect Ellie and keep her safe. It's what I've trained you for all these years."

I'd do it anyway, and he already made me promise this, but I tell him again. "I promise."

Kluge closes his eyes and the trace of a smile crosses his features. For a long time, he lies there, immobile. As I backstep to sneak out, he starts softly singing. I've never heard him sing before. Looking closer, I realize he's in another place now, not here in this hospital room with me.

"Where are you? I've waited. Do you see? ... For you, I'll surrender my entire life. Do you hear? ... How to find my way to you? For you, I'll light.... I love you eternally." Is he singing this for Lina, for his daughter, for his wife, Syntyche, or for all of them? Does it really matter?

The singing ends and he's deathly still. *Beeeeeeeeeep!* The heart monitor sends notice of his death.

Joe

Late Sunday morning, October 14

To say these past few days have royally sucked doesn't even come close. My hand is totally messed up, for which I'll probably miss the rest of football season, I got abducted then interrogated by Medusa, that was beyond infuriating, and I got roughed up by Anna's father, who's still insisting I kidnapped her and is hell-bent on putting the beat down on me. How am I going to explain away the black eye and busted lip he gave me? I sure as heck can't tell the truth, no one would believe me anyway since it's more convoluted than a fictional tale. And to top all that off, I got dragged into the basement of a homeless shelter to more than likely be framed for the killing of Markow, because who was the only one with his face showing for whatever surveillance cameras were in the area? Me. *Shit!* Either the police or the Solntsevskaya Bratva will be after me for that.

Yet all those problems combined are only the tip of the iceberg compared to the other gargantuan elephant on the dance floor. *Jeez, I'm sounding like Hannah.* After three long years of futilely searching for Lina, I have a lead! Someone saw her. And where am I? Stuck in the middle of Bumfuck Nowhere, tied to a chair. But if I wasn't tied up, would I be able to leave Anna now after what she's been through? After what I've been through with her?

Bloody hell! She's crashing and burning again. My head feels like it's going to bust into a million pieces. *Uh, oh!* Now I don't feel anything radiating from her at all, that's even worse. She's retreating down inside herself. If she keeps crawling deeper and deeper, will I be able to reach her?

"I want answers and I want them now, Carleton." Mr. Alkaev barrels into the room, holding an unconscious Anna in his arms. The serpent Medusa, who deigned to trail a few feet behind him, is nearly knocked flat when he forcefully kicks the door shut in her face. Her father then gently places Anna on the floor and methodically wraps the blanket tighter around her. "How long has she been getting these headaches?" he savagely yells.

I'd categorize them more as panic attacks, but arguing semantics with him at this point seems like a bad plan.

"Has she seen any doctors about the headaches?" he demands.

How the heck am I supposed to know that?

"If, as you claim, you aren't part of the kidnapping and you care about her at all, you'll start answering my questions."

It's the first thing he's thrown at me that has any sway. I do care about Anna, very much. *Think back, Joe, when did she start getting these panic attacks?* The first mind blowing explosion was just a few weeks into chemistry class, the next was the night she approached me at the Contented Cow Pub, then there was the first day of chemistry lab, the time I crossed paths with her while I was jogging, our second study session in the library when she looked at the photo of the Pyramiden playground, the first time I asked her to tell me about her mom, right after Ms. Mizuikova gave her Viktor's letter, upon hearing about The Fukushima nuclear disaster, on the ride home from the Eau Claire Post Office ... and the list goes on. "She's been having headaches since I first met her, and I don't know if she's seen any doctors," I grimly reply.

"Is she taking any medication?" Her father continues questioning me as he folds up his scarf and places it under her head.

"I don't know," I lamely answer. "Untie me and let me go to Anna. You've seen how I can help her." It's torture being so close and not able to reach out to her.

"How often does she get the headaches?"

Somehow, saying, *Whenever she's around me,* seems like a risky response. "Why don't you ask your pet snake Mizuikova these questions?"

Ominously, he steps over to me. "If Tatiana didn't need you right now, I'd beat the crap out of you. She's more important than either of us, and she's in grave danger. Her mother died from migraines like these, so keep the wiseass comments to yourself." He runs his fingers jerkily through his hair as he glances back at her. Anna's face is a sickly pale color. "Tatiana didn't get headaches before arriving here. What have you done to her, Carleton?" The viciousness in his voice is slightly less intense. He's scared he's going to lose her.

What have I done to her? I have no idea. Looking over at Anna, lying there, seemingly dead to this world, I'm crushed with searing regrets. I never should have gotten involved in her life. "I never wanted to hurt Anna. I didn't realize ..."

"Didn't realize! How could you not realize what you were doing to her? She's never needed anyone, never depended on anyone and never reached out to anyone since her mother died. She's cold, calculating and detached. It's how she's survived all these years."

Her father grabs me and starts shaking me, his face distorted with anger and worry. "What did you do?"

"I ... I sent her away when she asked for help," I confusedly mutter.

Momentarily, he's perplexed by my answer. "When did you first get involved in this kidnapping scheme?"

"I had nothing to do with that," I bitterly argue.

"Then how do you explain your fingerprints on the ransom note and the email from Markow? All the evidence points to you, Carleton."

The story of my life. He's going to rough me up no matter what I do, confess, try to explain (which he won't believe anyway) or stay silent. Well, I deserve what's coming to me, so I might as well try to explain. Maybe he knows something I don't, since his wife had these headaches too.

"I stayed away from Anna at first. I didn't want to get involved." I glance over at her again. *What have I done to you, Anna? I've failed you, just like I failed Lina. I never should have talked to you, I never should have kissed you on the sidewalk outside the pub, I never should have gotten involved in your life at all, you'd be far better off now.* "I ended up taking her out of a world where she was safe," I mutter without conscious thought that her father's still looming over me.

The blow he throws at my gut almost knocks the wind out of me, but it's nothing compared to my own internal penance. Having heard enough of my 'confession,' he zooms his attention back on Anna, throwing himself down beside her.

My memory starts dredging up more remnants of conversations with Anna, condemning my interactions with her and feeding my guilt. I can't sit here and do nothing. Fortunately, when the goons tied me up this time, they were careless and didn't notice me flexing my good wrist outward to create a gap in the ropes. I've been working to loosen the knot ever since. I pull at it more brusquely; this is taking too long. Bracing for the pain, I jerk my good hand free and gasp air in through my gritted teeth.

Anna's father is still slouched over her, unaware of my actions. As surreptitiously as possible, I free my legs, ease out of the chair and creep toward her. Awkwardly stretching out my good hand, I touch the bare skin of Anna's ankle that's sticking out from under the blanket. She so distant ... so unreachable. "I'm sorry, Anna."

The next thing I know, her father's on top of me, clenching my throat. I'm getting really sick of this repeating pattern. "I could've knocked you out just now," I rasp.

His grip loosens a bit. "Why didn't you?"

"Because you're right, Anna is more important than either of us."

His eyes narrow with suspicion. "Are you part of the Solntsevskaya Bratva or did the Kremlin send you?" he grills me.

"Can't you get it through your thick skull that I wasn't part of that?" I try pushing myself up, but I can't get enough leverage with only one hand. "Get off of me, Anna needs my help."

"Anna has never needed anyone," he angrily rebuffs me.

"After her mother's death, she needed you!" I bitterly attack back. "And you weren't there for her."

He grimaces and his hold on me loosens. He's nowhere near convinced he should let me anywhere near his daughter, yet he's worried what will happen if he takes me away from her. "At the homeless shelter, before I heard Tatiana scream, you told me she was downstairs. How did you know that if you weren't part of the kidnapping?"

"Because I felt her pain, and she feels mine." Even though it's doubtful he'll buy into it, I tell the truth.

Mr. Alkaev abruptly releases me and falls back onto his ass with a thud. The hard armor he wears to protect himself from the world is buckling. "How is this happening again?" he mourns with a sense of defeatism. He's not going to fight me anymore, at least not right now. I slide to Anna's side and soul-searchingly stare at her. There are no emotions running between us, none at all. Clasping her ice-cold hand, I call to her, "Anna, I'm here." She remains comatose. Tenderly cupping her face, I lean in and kiss her. Nothing. I can't connect with her! *Where are you Anna? How can I help you?*

Searching inward, I find my life-long mentor, the man with a shock of untidy white hair. *"Plato once wrote that every heart sings a song, incomplete, until another heart whispers back. Those who wish to sing always find a song, Joseph."* He gives me a reassuring wink, then counsels me to let the music to flow through me.

We've already established I can't hear the music, Mr. Einstein.

"Music in the soul can be heard by the universe, Joseph, it connects everything. Even when all else is gone, when language and coherent thought disappear, music still touches the soul. Follow your Lina, Joseph, she's been singing to you all your life. Her inconceivably tiny strings of energy underlie everything that happens."

I start urgently scanning all my musical memories linked to Lina. When the image of Kluge lying in the hospital bed before his death comes up, I stop. That's it! A universal melody that spans across space and time. Hesitantly, I start singing. "Where are you? I've waited. Do you see? ... For you, I'll surrender my entire life. Do you hear? ... How to find my way to you? For you, I'll light." I bend in close and whisper in her ear, "I sang for you, Anna. I sang strong."

Zing! A spark of energy zaps between us as if in greeting. Embracing her fully, I kiss her deeply, and we're whisked into that unexplainable plane of existence that's filled with strings of swirling, sparkling energy. This time, the strings around us are singing, and the indescribably beauty of this miraculous expanse of eternity is astounding.

I understand, Mr. Einstein. This link between Anna and me, it's not something we can control, it's not something we're doing or not doing that causes it to intensify or wane, it's the universe within us shifting and regrouping, changing and shaping. Evolving.

Anna opens her eyes, and I look lovingly into the beautiful, brown windows of her soul. Finally, I stop fighting this link and allow myself to accept this special entanglement of ours.

Chapter 42

One interpretation of the Jewish mystical tradition of Kabbalah. In the beginning, there was only the darkness and silence, Ein Sof, the source of life. And then, from the heart of this darkness, this world of a thousand, thousand things, there emerged a great ray of light filled with song. But an accident broke the vessels containing the light and music of the world, and the wholeness of existence was splintered and scattered into a thousand, thousand fragments, which fell into all people. The wholeness of the world remains there yet, deeply hidden ... but some have the capacity to see this unseen source of life and make it detectable once again, restoring the innate connectedness of our world.

Tatiana
Noon on Sunday, October 14

Joe! I force open my eyes to see him staring down at me, and I gaze back in amazement. Finally, I've attained some of the understanding I've been seeking. The universe is alive and responsive!

Wham! I'm cruelly ejected from this glorious glimpse of an infinite, interactive realm to crash-land in the dim, dreary confines of a tiny, finite room. My father's worried face fills my vision. Twisting, I see Joe scrambling to sit up. "What the hell did you do that for?" he explodes.

“I kept shouting at you, but you were totally unresponsive,” my father yells back. “All the color was draining from your face. I can’t risk having both of you unconscious because I have no idea how that would affect Tatiana.”

Waaaahhhhh, Waaaahhhhh, Waaaahhhhh. The sound of a siren invades our cramped space, filling it with a sense of impending danger. As the wailing noise intensifies, a hound in the surrounding woods starts howling with it, creating an even more fractious warning. *Waaahhhhhooooo.* All of us freeze, seemingly held captive, as the earsplitting noise reaches its peak. When it speeds on past, we let out a collective sigh of relief.

“I have to take you somewhere safer, Tatiana. Too many people know you’re here.” My father’s rote reaction rushes out of his mouth.

I sit up and shimmy my back against the wall. “What you need to do, Father, is let me make my own decisions.” *It’s long past time for that, and whether you decide to accept it or not, you’re done dictating my fate, given that I’m done with your rigid rules and controlling authoritarianism.*

“I’ll let you peruse the list of places I readied, and we can jointly choose where you should go.” This is his version of allowing me the freedom to exercise my own judgment. I know from past experience, it’s a very false illusion of freedom.

“No. I’m staying here at Carleton.” In an odd sort of way, this place has grown on me, and I’m starting to *feel* genuinely happy here, or at least a resemblance of happy at times. I’m not willing to give that up, no matter the risk.

“It’s too dangerous, Tatiana. Too many people know you’re here.” My father reiterates this with his usual dictatorial authority.

“What people, Father? Markow was working alone.” *Okay, Joe, you can jump in anytime to corroborate this and refute your involvement in my kidnapping.*

My father's face hardens. "Carleton was part of this scheme, and we still don't know if there were others. You have to relocate." His steel look of determination solidifies. His mind is already made up, and nothing I say is going to change that.

Well, my mind is made up too. *"I'm* remaining here, and *you* are leaving, along with all your security forces, including Ms. Mizuikova. I refuse to live under your constant surveillance anymore." If he won't do this, then with Joe's help, I'll disappear from him forever. "Markow told me he was going to use the ransom money to buy his way out of the Solntsevskaya Bratva. He wasn't working with them, and Joe had nothing to do with my abduction." Wordlessly, I signal for him to confirm this, but he doesn't. Why not?

Brinnng, brinnng. My father's cell phone rings through the barriers of our impasse. Ripping it from his pocket, extremely ill-tempered due to my noncompliance, he shouts into it. "What do you want? … Yes? … You're absolutely certain Markow was acting alone?" There's a pause, then he switches to a hushed tone. My father has never really mastered this, and Joe and I easily overhear what he's saying. "That song you sang long ago, next to my wife Katya's casket, what are the words? … I know Katya was your daughter. Just tell me what the words to that song are." My father's face contorts with resentment. "She's long dead, continuing to blame me won't change that. We need to focus on Tatiana's safety, and it's critically important I know the words to that song." Abruptly, my father shoves the phone at me.

"Are you all right, Tatiana?" Grandpapa asks, his voice sprinkled with concern.

"I think so, Grandpapa." I don't really know how to answer at the moment with all that's happened.

"Why is your father asking about that song?" Deep pain and a bit of anxiety are mixed into his question.

"What song?" I wasn't at my mama's funeral.

"So it has nothing to do with your safety?"

"I don't know, Grandpapa. Sing the song for me," I suggest.

He breathes a sorrowful sigh, then starts singing. *"Where are you? I've waited ... Do you see? For you, I'll surrender my entire life ... Do you hear? How to find my way to you? For you, I'll light."* I hear tears in his words.

There's a certain sweetness to the song. It sounds so familiar, yet pinning down where and when I heard it is elusive. "I think I've heard it, Grandpapa. Maybe you sang it to me before, but I can't be sure." Meanwhile, my father's anxiously tapping his foot right next to me. The distraction isn't helping. I look up. "Why are you asking about this song, Father?"

"I'm trying to figure out how Carleton ties into all of this. He sang that song to you just now, when you were passed out. It's far too implausible to be pure coincidence. That song was written a long time ago and is no longer popular. So why would he be singing it? I'm coming to the conclusion that your grandfather hired Carleton to keep watch over you. It's exactly the type of twisted thing he would do."

What? After the initial shock recedes, I stop and consider it. Joe's knowledge of Grandpapa's favorite expressions, his knowing that Alyonka chocolates are my favorite, his refusal to talk about himself, his reluctance to help me escape and his misgivings about kissing me. If he's working for my grandpapa, all that makes sense. "Why did you sing that song to me, Joe?" I half ask and half accuse him.

Swirls of trepidation and reticence swim in his Siberian blue eyes. "I'd never do anything to hurt you, Anna," he evasively answers, like he always does when I press him for personal information.

The ugly bitterness of being duped starts churning in my gut, almost choking me with its corrosive acid reflux as it invades my throat and mouth. "*This* is what you've been hiding from me, that you were hired by my grandpapa to babysit me!" More nasty feelings of betrayal claw at my heart, and I scream into the phone, "Grandpapa, did you send Joe here?" He, too, declines to answer this question. "Did you?"

I interpret his continued heavy silence to mean *yes*. Fully boxed up and cornered into a livid little hole, I throw the phone across the room, then whip my fury at Joe. "I actually believed you cared about me when all along it's been a ruse. You've only been spying on me for my grandfather." How could I allow this to happen? How could I have been so stupid?

Joe hangs his head, avoiding my scathing gaze.

"If you're willing to keep a GPS tracking device on you and have a small security detail, I'll allow you to remain at Carleton," my father cuts in. "Now that I know Markow was working alone and Carleton's working for your grandfather, the situation has changed."

I'll say the situation has changed. "I don't want to be at Carleton anymore," I belligerently gripe. *'But I do want to stay at Carleton,'* the bit of my heart that still yearns for Joe cries out.

"The tracking device is barely two centimeters in diameter and just over a centimeter thick, so it would be completely inconspicuous." My father plows ahead with his change in plans without even listening to me, once again totally unconcerned about what I think.

"I don't want to stay here!" I shout directly at him. "And I don't want to have anything to do with Joe ever again."

"We'll fly out tomorrow, then." Seamlessly, he switches gears, pleased at the way things are turning out.

"Why didn't you tell me you're working for my grandfather, Joe?" Their joint betrayal of trust is splitting apart my soul like the break up of Siberian ice at the start of spring. I'm left dangling alone, pathetically grieving the loss of a future shared life with Joe that I had foolishly imagined was possible. Add in my father's total indifference to my feelings, and I'm left with a destabilizing, debilitating angst. "Why did you keep this from me, Joe?"

He continues concentrating his attention on a certain section of the floor in front of him while his feelings zing all over the place. Sorrow, disappointment and ricocheting guilt.

"Get out, Carleton," my father orders with unabated delight.

Joe heads for the door, and a different undercurrent of emotions passes between us. Sadness, wistfulness ... and something else. He's still hiding something from me.

"I'll have Ms. Mizuikova go to your room, gather your things and stop by the registrar's office to disenroll you," my father rambles on as Joe hurries away.

What else is he keeping from me? There was a smidgen of something Joe was hastily tucking away before I could notice, something that doesn't fit, a sort of melancholy déjà vu. *T'fu!* What a mass of tangled webs has been spun behind scenes that I wasn't privy to. That's going to change. I'm fed up with being a pawn, and I'm not going to stop looking into this until I find out the entirety of the disjointed mess that was orchestrated behind my back without my consent.

"I'll arrange for a hotel room for tonight. Be ready, Tatiana, we'll be on the first flight out in the morning."

Hotel for the night, flight out tomorrow? *Pizdets.* While I've been internally fuming and contemplating my course of action, my father's been steamrolling along with his. Quickly, I put an end to that.

"I'm not going anywhere with you, Father!"

Chapter 43

Joe, age sixteen

Ever since the attack by that Russian gang member, Kluge has been testing me. Feeding me bits of information, then purposely inciting me. Deliberately prodding, provoking and antagonizing me, then analyzing my reactions. He's been erratic, irascible and conflicted. After a month of this, I've had enough. His crazy notions of conspiracies are getting out of control.

Walking into his back study, I start telling him to bug off. "I'm tired of all your annoying—" With a speed I didn't think he still had, Kluge shoves me up against the wall and pins me there. He's weak now, and I could easily break his hold on me, but I don't. After glaring for a long time, he finally reaches some sort of conclusion and releases me. "Sit down, Joseph, I have a lot I need to tell you."

I reluctantly take a seat on the far end of his old leather couch.

"When I was ten years old," he begins. *This could take a long time, considering what an old geezer you are. I'll give you ten minutes.* "There was an early morning knock on the door of our apartment. It was the NKVD, Stalin's secret police. They had an arrest warrant for my father, accusing him of crimes against the state. The entire day they spent ransacking our apartment, searching every closet, cupboard, drawer and shelf while we huddled in a corner. Once they finished rifling through our belongings, they boxed up all the papers, boarded off the bedrooms and hauled my father away. After torturing him for days, they got their 'confession' and he was executed. The only crime my father ever committed was speaking out against Stalin's ideas." The confused heartache of a little boy is still present in Kluge's words.

"What happened to you and the rest of your family?" I ask.

"My mother, older brother and I were deemed enemies of the state and sent to separate prisons. Much later, I found out my mother died about a year after being arrested. My brother and I were held in an institution for underage offenders, where he contracted tuberculosis. He didn't survive." Kluge looks hard into my eyes to emphasize his next words. "I was left all alone in this world and forced to make a choice, give up and die or fight for my life. I chose to fight."

"How did you get out of prison?" This tale is more interesting than I thought it would be.

"A few months after my brother's death, there was a fire, and in the confusion, I managed to escape. For weeks, I kept running and hiding, never staying anywhere long. Eventually, I stumbled upon an orphanage. I worked hard there, studied harder and got perfect grades. When Stalin died, I applied and gained entrance into Moscow State University, graduated with a degree in engineering and went to work at the Mir diamond mine."

So that's his connection to the Silver Bear diamonds.

Kluge gets a faraway expression, and a wisp of a smile fleetingly crosses his face. "I fell in love and married Syntyche Dobrow. She was an amazing woman, strong-willed, smart, impulsive and extremely outspoken. A bit like you, Joseph."

That's the first compliment he's ever said to me.

"During Khrushchev's rule, Syntyche's differing political views were overlooked, but when Brezhnev took over, things changed dramatically. Her cutting accusations, combined with her vigorous support of Aleksandr Solzhenitsyn's literature, attracted the wrong type of attention, and the KGB started monitoring us. For a while, Syntyche managed to keep her opinions to herself, but shortly after the birth of our daughter, she started challenging the ruling party again, wanting a better world for our child. I was scared for our future and devised a brilliant defection plan, but it was intricate and required time. In the midst of my preparations, Syntyche was diagnosed with terminal cancer." Kluge sinks back into his chair, weighed down by the oppressive memories.

Wow, his life's been way shittier than mine.

"Before she died, Syntyche made me promise to take our daughter somewhere safe, somewhere she could grow up free to speak her mind and follow her spirit. Mere hours after the funeral, I grabbed the satchel of Silver Bear diamonds I stole, gathered our daughter up in my arms and disappeared into the night. With the aid of a few good friends, I covertly crossed the country and eventually made it to the United States." He pauses for a long intermission.

"What else do you need to tell me?" I'm starting to get impatient.

"We were never really free, Joseph. Having lived through Stalin's Great Terror, where over one and a half million people were executed or imprisoned for defying the regime, I was intimately aware of the evil they were capable of. In order to keep us safe, I continuously changed our identities and moved from state to state. It was a horrible way to live, constantly looking over my shoulder in fear." His eyes glaze over as he pulls a photo out of his shirt pocket and wistfully stares at it.

I'm tempted to get up and look, but I doubt he'd let me see it.

"I never told my daughter anything about her past ... and she grew up without ever knowing where she came from, what her birth name was or anything about her mother, but I kept my promise. She grew up free." With deep reverence, he places the photo back in his pocket. "You would have been proud of her, Syntyche." He pauses and stays silent for a long time, lost in his memories.

Okay, enough morbid reminiscing. "Decades have gone by and no one's ever found you, so why continue hiding everything from your daughter and Lina? And why don't you ever go see them? They only live a few miles from here."

"My daughter died many years ago," he sharply rebukes me.

Lina has a stepmom? It's strange she hasn't mentioned this to me, we tell each other everything. "Does Lina know her birth mother died a long time ago?"

"Stop interrupting me, Joseph," he fumes. "After graduating secondary school, my daughter started college, and that's when she met that condescending SOB who got her pregnant. Against my wishes, she dropped out of school and married him. For months, I pleaded with her, telling her we could move anywhere she wanted if she would leave him, but she always refused. Even after he came home one night in a drunken rage and hit her, she still spurned my help and remained with him." Kluge shakes his head, not able to comprehend her actions.

"Was he ever arrested for that?" I ask.

Kluge directs his residual anger at me. "Keep your mouth shut and let me finish."

"Fine," I spit back. "Finish already."

"As much as I wanted to kill the bastard, I couldn't because it would have devastated my daughter. Instead, late one night, I grabbed my pistol and waited for him outside the bar he frequented. When he came out, I shoved the barrel in his mouth and threatened that if he ever hit my daughter again, he was a dead man. The stupid ass went straightaway and told her what I did. She was furious and never spoke to me again, but it was worth it because he never hit her again."

"Is there an end to this?" While his long saga is intriguing, I'm getting tired of sitting through it.

Kluge scowls, then continues with the condensed version. "A few months after Ellie's birth, my daughter was killed in a car accident."

Weird, Lina's stepmom died in a car accident too. Is her father capable of that level of premeditated violence?

"Her husband took the baby and moved to Iowa. A year later, he remarried and they all moved to Ohio. Two years after that, they moved to Wisconsin. Three years after that, they settled in Hibbing ..." His voice trails off.

"And?" I frustratingly ask.

"And that's when you met my granddaughter, Ellie."

Joe
Early Sunday afternoon, October 14

I leave without even saying *goodbye* to Anna. I want so badly to tell her I didn't betray her, that I wasn't working behind her back for her grandfather and I would never use her emotions to gain advantage over her. But if I do that, her father will search further into my past, and I can't let that happen. I promised Kluge I'd keep Lina safe and never reveal anything about their past to anyone. I won't break that vow, even though it's highly unlikely after all these years that Mr. Alkaev could uncover Kluge's hidden past; he changed his name many times over the decades. But Kluge's words echo deep in my soul. *Are you willing to risk Ellie's life to prove you're right, Joseph?*

I have to pass by the serpent Medusa to get out of this decrepit old house. Her eyes are glistening with malice as she hisses at me. Bolting down the porch steps, I take a thorough glance around, and the immediate circumstances of my situation slam into me—miles to walk, my hand's killing me, my mind's hashing through memories like a meat grinder, I'm ravenously hungry and I don't have my cell phone. *Just freaking great!* Swearing up a storm, I begin the lengthy journey back to campus.

What should I do when I finally get there? I ought to go get my hand looked at, yet if I don't head straight to the homeless shelter to ask that guy about Lina, the only lead I've ever had may go completely cold. However, showing my face there is dicey, especially if anyone's discovered the blood in the basement. Also, if I do happen to find out anything that helps me locate Lina, and Anna's father is still having me followed, I might be leading him and Anna's grandfather right to her. My gut tells me they're connected to her somehow, side characters in the extended Kluge saga. Yet I can't not go, I can't give up on Lina.

A few miles of plodding along the road brings me to a path into the arboretum. As I'm about to step onto it, the infamous black sedan pulls over a few yards in front of me, and Anna emerges. She's wearing gray sweatpants that are too short, a blue sweatshirt with an embroidered loon on the front that's at least two sizes too large and running shoes. I'm guessing these clothes are Medusa's. She marches toward me with a roiling mixture of tempestuous emotions.

"What do you want from me now, Anna, a confession signed in blood?" I angrily ask, wanting to get rid of her as fast as possible because I don't want to hurt her any more than I already have. My vow to Kluge prevents me from telling her the truth and explaining this whole mixed-up affair, so there's no way to give her the answers she's looking for.

"I'm fed up with everyone making decisions for me without consulting me!" Taking a page out of Hannah's book, she pokes me in the shoulder. "I'm sick of all the secrets in my life, the ones my father and grandpapa keep from me and the ones *you* are still keeping from me, and I'm especially upset by the cat-and-mouse game you play with me. You're still hiding something, I can feel it. Fess up or get ready for battle." She balls her fists up as she unloads this ultimatum on me.

"I don't think you're physically capable of stepping into the boxing ring with me, Anna. You should be getting medical attention after the ordeal you've been through." I attempt to divert her, although, I am genuinely concerned about her condition.

"I'm fine," she tersely snips. "You're the one that needs medical attention." She points to my wickedly swollen hand.

"I'm fine," I clip back. "Or I will be as soon as you get the hell away from me and crawl back to your daddy." *Please go!* I dart past her onto the path. There's only room to walk single file.

"Where do you come off as being the one who's been misused in this bloody mess? I'm the one who has been kidnapped, manipulated and deceived. I trusted you." Her voice blisters with hurt feelings.

I'm hopelessly caught, pulled by Lina at one end and Anna at the other in a hideous tug-of-war. "Life isn't always about trust, Anna. Sometimes it's about promises made and honoring them," I irritably tell her over my shoulder.

"Not all promises are honorable, Joe, and I deserve to know. Tell me what you and my grandpapa are keeping from me."

I can't do that, Anna. Shit, if she continues yanking my rope this hard, I'm totally screwed. I don't have the heart to keep this sham up much longer. Why is my life always so damn complicated? "The matter doesn't concern you," I mumble, needing to repress the alluring feelings that are drawing me to her and may ultimately lead me into spilling my guts.

"You're working for my grandpapa to spy on me. That directly concerns me."

She has a valid point ... if it were true. "Doesn't matter anymore, Anna. You're leaving Carleton, so you won't be seeing me again. Get the hell out of here!"

Somehow, she manages to zoom in front of me, forcing me to a halt. "Answer this, and then you'll never have to bother with me again. When you kissed me outside the pub, were you only doing that to manipulate me and make it easier for you to do your job?"

She's just handed me the perfect escape route, if only I could stomach it. Saying *yes* would cement her misconception that I'm working for her grandfather, nonetheless, I can't do that to her; I can't emotionally destroy her. Even though she's standing tall and rigid, presenting a strong front, underneath she's exhausted, scared and vulnerable. "No, Anna, I wasn't deceiving you. I honestly care about you."

Her composure wavers. She's not sure what to believe anymore, and she has no one to turn to. Impulsively, she wipes at her eyes, perhaps hoping to brush away the confusion and see clearly what lies underneath. I remain motionless, even though I want to flee the intensity of her feelings that are pouring into me. Finally, she comes to some sort of conclusion. "My real name is Tatiana Pavlovna Novikova, and I came to Carleton in order to avoid the perils against me in Russia."

Anna's spilling her secrets? *No.* I can't get suckered back into this mess. Her father and grandfather will ensure she stays safe, she doesn't need me in her life. I scoot around her, resuming the long trek back to campus.

Of course, Anna treks behind me. "My grandpapa is an influential oligarch and an outspoken critic of the Russian regime. After the collapse of the Soviet Union, he took advantage of his connections and political savvy to make a great deal of money which he then used to influence Russian politics. Unfortunately, when Putin took over things changed for him. To stay alive, he fled Russia and started publicly denouncing the Kremlin from afar. Due to that he's in their crosshairs, and according to my father, the scope is wide enough to encompass me." She states this matter-of-factly.

How many years has she been living with all this dangling over her head? *Don't let your feelings take over, Joe, just keep walking.* A whiff of wind sends the scent of lilacs my way. Lilacs were my mom's favorite flower, not only because of their sweet fragrance, but also because of what they symbolize. Renewal. But lilacs bloom in May, not October. This fragrance must be from the shampoo Anna used to wash her hair, which is still damp. It's a huge improvement from how she smelled just a bit ago.

"Why Carleton?" I demand. "There are thousands of other liberal arts colleges in the US to choose from. Why did you have to come here?"

"Carleton has over two hundred international students, so I don't attract undo attention, and my father insisted I go to a small school, not one in a large metropolitan area. In addition, Carleton offers over a hundred music classes."

Music! *Absolutely not. Don't let your memories take you there, Joe. Think facts.* The first game of football was played between Rutgers and Princeton on November 6, 1869. It was developed by integrating two sports, rugby and soccer. Integrating! Anna and I are integrated. How deep is our bond and how long will it continue after we're permanently separated? Will I ever be able to live a semblance of a normal life? Can this entanglement between us be broken? *Back to facts, Joe, relevant facts.* Anna's admissions about her grandfather explain some things, yet they don't give any clues concerning the strange ties her family seems to have with Kluge. Why does Anna know verses of the poem Kluge's wife wrote for him right before she died? Why did her grandfather sing the same song at Anna's mother's funeral that Kluge sang on his deathbed? I can't shake the odd notion that her presence here is purposeful. Why didn't her grandfather deny hiring me? Could Anna possibly be working with him to track down Kluge's diamonds? Maybe I'm the one who's being intricately duped and tricked. It would be quite the acting job, but I wouldn't put it past her.

Halting and turning to face Anna, I bluntly confront her. "Where did you learn the words to that poem?" Maybe there's a perfectly reasonable explanation.

"What poem?" She looks at me askance.

" 'Others will never know, how you reached out to me. You crossed the cosmos, and stretched eternity.' You said something very similar to that when you regained consciousness in the basement of the homeless shelter." Restlessly, I shuffle my feet as I anxiously wait for her answer.

"Where did *you* learn the words to that poem?" She cleverly turns it back on me.

"Would your grandfather use you to get to me, Anna?" I ask, hoping she'll negate these ludicrous conspiracy notions Kluge planted in my head and spell out her exact involvement in all of this.

"For what purpose, Joe?" She skillfully holds her ground.

"To find something he lost a long time ago," I supposition to myself as much as to her. *Jeez!* She's shocked by my reply. There is a connection. "Did your grandfather teach you that poem?"

"Why is this so important to you?" She's searching my face for some clue. A blaring car horn causes both of us to jerk, looking over we see a small pickup truck fly past, barely missing the parked sedan Anna arrived in. The front window of the sedan rolls down, and one of the guards leans out, signaling her to return. Anna signals back by whipping a flat hand a few inches from her neck in a cutting movement. Wisely, he rolls the window back up.

"What other verses do you remember, Anna?" There's still a slight chance the few lines she recited are just pure happenstance. I hold tightly to this slim possibility, reason being, it's hard to accept the alternative. Her grandfather is hunting down anyone connected to Kluge in search of the diamonds, namely me and Lina.

"I didn't memorize it, Joe. He gave it to me a long time ago when I was a little girl, right before he fled to England." She just confirmed she learned this poem from her grandfather. Did she admit it purposely or inadvertently?

"Then tell me whatever you remember." An isolated squall stirs up a bunch of fallen leaves, momentarily distracting Anna. I try guiding her along. " 'Sing for me and I'll be there, if you sing me strong. Sing when all the day is dark, and every star has gone. Sing when all the air is cold ...' " I pause, seeing if she'll fill in the blanks.

"I already told you I don't remember," she annoyingly states.

" 'And home's a far-off spot,' " I finish the line for her. " 'Sing for me when others won't ...' "

" 'Because they have long forgot,' " she softly completes the verse, her mood becoming contemplative. A lone monarch butterfly flutters around us for a few moments, then flits away, one of the last to leave for its three-thousand-mile journey south to the Sierra Madre mountains in Mexico. Anna watches it with keen interest, then refocuses on me. "Recite the rest of the poem to me, Joe."

Since she hasn't been cooperative in telling the lines to me, there's no use irritating her. Well, it's going to be out of order because I began by quoting the verses she spoke in the basement, which were near the end of the poem. I shrug my shoulders, shouldn't make much of a difference.

" 'Sing very strong. And shine, shine, shine with the song. Only you and I will know, this spooky entanglement of ours.' " Anna flinches when I mention *spooky entanglement*. The weird coincidence spooks me, too, and I shiver like I used to when I was a kid deep in the Minnesota woods gathered around a campfire with my siblings telling ghost stories. Hesitantly, I pick up where I left off. " 'That holds our souls together, through space and time and stars. Only the two—' "

" 'Only the two of us will know this secret,' " she cuts in, shaken by the parallel implications to our lives.

A few strands of loose hair fall across her face, and I reach out to push them behind her ear. "Finish the last line of the poem, Anna."

"There wasn't any more," she adamantly states.

"I thought you didn't remember it very well. How come you're so sure of that?" It doesn't make any sense that her grandfather told her this entire poem and then left off the last line.

"The ending was special. 'Only the two of us,' just Grandpapa and me knew our secret," she says with a sense of finality. "Now tell me why this poem is so important to you?"

It's important, Anna, because it now seems that the secret "only the two of us will know" is known by more than two people. More significant, though, is who shouldn't have discovered this secret, Kluge or Anna's grandfather? "The verses to that poem aren't published anywhere, and I can't figure out how your grandfather knows them."

"Why would you care about that?"

'Cause I need to find out if your grandfather is a danger to Lina. "The person who originally wrote that poem died shortly afterward, and the person she wrote it to is dead now as well. There weren't any other copies, or at least I assumed there weren't. Obviously, I was wrong."

She screws her eyes up in genuine confusion. "What are you implying?"

Is she really not privy to her family's secrets, or is she deftly performing her part to garner information out of me? The years spent with Kluge have affected me more than I want to admit. I wouldn't ever have had such outlandish plot twists in my head if he hadn't put them there. *Mr. Einstein, help me out here.*

Einstein is lying on his deathbed in Princeton Hospital. He asks only for his glasses and his latest equations. "I want to know God's thoughts—the rest are mere details," he explains.

Thanks for the advice. I interpret the parable for Anna. "The last hours of Albert Einstein's life were spent trying to decipher the mysteries of the universe. He ended up dying isolated from the scientific community and ridiculed by his peers. Like us, Anna, Einstein was bedeviled by secrets."

"I have no intention of wasting precious brain cells helping you decipher Albert Einstein's secrets."

Wow. She totally spaced whose secrets I'm referring to. Or did she? "Then let's place ourselves within the quandaries of the exotic physicists Alice and Bob. In their quantum world—"

"I don't care about Alice and Bob either!"

"But you care about your grandfather. What did he do before he became an oligarch?" The words are out of me before I can recall them. *Subtle, Joe, real subtle.* You needed to get her more riled up and off balance before you hit with that.

"He was an officer in the Foreign Intelligence Service of the KGB," she rattles right off.

For real? Shit! With those type of job skills and connections, he quite likely *is* searching for the diamonds. For Lina's sake, I have to know with one hundred percent certainty. Time to meticulously scour my personal database. Thinking all the way back to the start of this convoluted chess match of wits, my first encounter with Kluge, I start analyzing.

Anna's saying something, but I block her out in order to methodically connect all the dots ... no matter which pathway I follow, they lead to the same endpoint. Kluge and her grandfather knew each other. What was their relationship? Was it hunter/prey, or were they coconspirators?

"Are you listening to me?" Anna grabs me by the shoulders, temporarily interrupting my concentration. Twisting away from her, I attempt to refocus. What should I do? It's too late to untangle myself from this mess. I have to determine how to become a more powerful player so I can decidedly alter the outcome of this deadly game.

"Joe!" Anna screams right in my face, her rising fear racing across me like a fleeing rabbit. She's just a pawn in this game, like me, I can feel it; I've felt it all along, only I haven't wanted to trust my feelings or follow where they've been taking me. Whatever scheme her grandfather's contrived, he's selfishly using Anna to his advantage. She's not intentionally acting a role, only inadvertently, if at all.

Damn it! She shouldn't be involved in any of this, but what can I do about it? I'm intent on not fouling up her life any further, which is certain to happen if I stay in contact with her. Yet, maybe I can fill her in a little on what I know is going on, as a parting gift. However, there's a danger I may let too many details slip, that often seems to happen when I'm around Anna. Then something bad could happen to Lina. Still, if I don't help Anna—

"Jooooe!"

Anna's right, her life shouldn't be controlled by others who don't inform her of the consequences or ask her consent. She deserves to know the truth.

"I had nothing to do with your kidnapping, Anna, and I wasn't hired to keep you safe. I don't work for your grandfather or your father or Markow or the Solntsevskaya Bratva or anyone else in Russia."

Chapter 44

As long as I have any choice in the matter, I shall live only in a country where civil liberty, tolerance and equality of all citizens before the law prevail. The right to search for truth implies also a duty; one must not conceal any part of what one has recognized to be true.

Albert Einstein

Tatiana, age eighteen

"It doesn't have to be anything valuable, Tatiana, just something special to you."

Really Viktor. I roll my eyes, an expression I've learned by watching others is meant to display annoyance. It sometimes works to get people to stop doing what they are doing, and I want Viktor to stop. He has been endlessly badgering me for days to give him something emotionally dear to me that he could carry in his pocket when he ships out with his unit for his tour in Chechnya.

"I don't have anything emotionally special to me, Viktor. The answer won't change by you repeatedly asking it."

"Then just give me anything of yours that's small enough so I can bring it with me," he persists.

A brilliant idea pops into my head. I can unload the last one of those bloody diamonds on him. They've brought me nothing but misfortune whenever I use them. It will be good riddance.

The next evening, ten hours before he's due to ship out, I give him the diamond which I sealed in a small jewelry box, and instruct him not to open it until the following night. I don't want him refusing to accept it because he thinks it's too valuable. And even though I'm not at all sentimental, I don't want him to be pining away believing I didn't care enough to give him something as a remembrance. Hopefully, the diamond will fill that bill.

Well, it didn't turn out at all the way I wanted. The blasted diamond got him killed.

Tatiana
Sunday afternoon, October 14

"You don't work for my grandpapa or anyone else in Russia?" This makes no sense. "Then why didn't either of you deny it? How do *you* know the exact words to the song he sang at my mama's funeral as well as the words to the poem that he gave me, which you claim wasn't published?" *And what were you thinking about so intently just now when you blanked out on me?*

Joe chooses this moment to skirt around me and resume hoofing it back toward campus. *Oh no, you don't.* I quickly latch on to the hood of his sweatshirt. "You can't just walk away after what you've said to me. Tell me what the last line of that poem is." *Really, Tatiana? That's what you have to know? What about who are you really working for, Joe, and if you're not working for anyone then how do you know so many personal things about my family?*

He spins around, unclenches my fingers from his hood one by one and then replies, " 'We are immortal.' "

Is he being philosophical or is that the last line? A strong sense of connectedness lunges at me, whooshing the breath out of my lungs.

Tilting my head up for air, I spy a hawk skillfully circling on the currents. "Wish I had that type of freedom," I mutter. The raptor zeros in on its prey, like a dart headed for the bull's-eye, and flies away with its prize. Wish I could do that too.

"Charles Dickens once wrote 'every human creature is constituted to be a profound secret and mystery to every other, and every beating heart is, in some of its imaginings, a secret to the heart nearest it,' " Joe gravely recites. "His vision rings true for your grandfather, as well as for us, Anna. Can you remember any of his friends or business associates from around the time he gave you that poem?"

What's his angle? *Be cautious, Tatiana.* While he claims he's not working for anyone from Russia, maybe he's currently employed by the US government and is some sort of double agent. *T'fu! My father's put such crazy notions in my head.* Still, I should tell Joe to bugger off and be done with it because my life has been turned upside down since we met ... but I don't want him to bugger off. I like having him in my life. As I'm wrestling with what to do, that unique form of 'Joe' guilt is rolling off him in torrents. It's possible he really is an innocent bystander in all this. Impulsively, I do what I never did before I met him—I make a spontaneous decision based on a gut feeling. "I don't know anything about my grandpapa's work, Joe. He never talked to me about it."

"You must have heard something over all these years," he insists.

"I wasn't allowed to talk to my grandpapa anymore after he fled to England. My father forbade it." *What a terrible thing that was for me, Father.* "I'm so sick and tired of my father controlling every bloody thing in my life!" It feels incredibly good to scream this out loud at the top of my lungs.

"He forbid you from talking to your own grandfather?" Joe's totally aghast at this.

"Yes, my father was furious with Grandpapa for speaking out

against the Kremlin and putting our entire family in danger. He was also fed up with Grandpapa constantly blaming him for my mama's death. After an epic argument, my father cut off all communication with him, swearing he'd never talk to his grandchildren again." This is sort of fun, flaunting my father's stringent rule about not revealing personal information to anyone. It's rather uplifting and freeing to tattle about these family flaws.

Out of the corner of my eye, I notice the sedan doors open. The two guards get out and start approaching us. It almost seems as if my father has planted eyes and ears out here in the woods, part of an all reaching network that notifies him if I'm revealing anything I shouldn't. *Well, you did scream loud enough to wake the dead, saying you were tired of your father controlling your life, Tatiana.* When they're within hearing distance, I preemptively cut them off before they can say a word. "I don't care what my father instructed you to do, I'm not leaving until *I* am ready to. Get back in the car." Neither one retreats a step, however, they don't come any closer either.

"Your father and grandfather appear to be on talking terms now," Joe points out in a hushed voice using English, assuming the guards aren't fluent in it.

I'm not taking that chance. Linking my arm through his, I lead him off a ways where we can't be overheard. It's too cumbersome for me to have an in-depth conversation in English, so I revert back to Russian. "After I was attacked two years ago, my father and grandpapa had a reconciliation of sorts to coordinate issues regarding my safety." Continuing to defy the excessive restrictions my father placed on me is giving me a runner's high. "In addition, my grandfather found ways to get messages to me."

"Through who? Can you remember what they looked like, what their names were or anything about them?" He seems overly eager to find this out. "I need to uncover the connection between your grandfather and that poem, Anna."

Not before I find out what you're still keeping from me, you don't. "If I agree to help you find out what you want to know, do you *promise* to answer my questions first?" I want his verbal confirmation on this, because binding promises appear to work on him.

"I'm done making promises, Anna."

"What a convenient time for you to turn over a new leaf," I sarcastically comment.

"You're right, it is a good time for leaf-peeping." He points to the trees next to us. "The sugar maples are turning orange." He directs my attention to some bushes by the guards. "The sumacs are turning red." He motions to some trees next to a small stream. "And the *whomping* willows are turning golden."

Joe's wordplay doesn't amuse me or distract me. "I'm tired of all the secrets between us, Joe. If you're not working for my grandpapa or anyone else in Russia, why do know so much about me and why do you keep pushing me away?"

"It's better this way," he flatly states. "Remember what I said to you after our first kiss? It hasn't changed. I can never be that special person to you, Anna, and you deserve someone who can love you with their whole heart. I can't do that, and if we tried to be together, I would only end up hurting you."

"News flash, Joe, you've already hurt me," I lash out. "And it's not for you to unilaterally decide what is 'better for me.' " *Ofiyet'!* Why must his emotions affect me so strongly? Staring at the hard set of his shoulders and feeling the raw mix of cutting pain scraping within him, I'm conflicted, furious with him yet not ready to give up an 'us.' Closing my eyes, I think way back and picture my grandpapa kneeling in front of me, handing me the folded piece of paper with the poem on it. Meticulously, I think through the entire encounter, and after a second run-through, I recall something that might be significant. "Grandpapa told me that a close friend wrote the poem for him."

"Keep your eyes closed, Anna." Joe's excitement jumps into me. "Stay in that moment and think through every part of it with me. Was it day or night?"

"Day." I can see the sun shining through our back window.

"Were there other people there?"

"No." My brothers were upstairs and father was in the room he sometimes used as an office.

"Was it summer or winter? Describe every detail."

"It was fall, and Grandpapa was wearing his heavy overcoat. We were by the back entryway and he was kneeling so we were face-to-face. His breath smelled like cigars. Grandpapa had come to say goodbye because he was leaving for London. He was glad we had a few moments to talk privately and told me I needed to wait for him—to wait very hard." I choke on the words, only now realizing how deeply this parting affected him, and ultimately me. "Grandpapa started to cry, and I was so confused. I didn't understand why he couldn't come back anymore and had no idea how to wait very hard." Taking in a huge gulp of air, I try to steady myself. "He reached into the inner pocket of his coat, pulled out the folded piece of paper, placed it in my hands and said that someone very dear to him had to go away and wrote this for him."

A high-pitched shrill pierces the air, breaking my concentration, and my eyes pop open. The entryway of my childhood home framing Grandpapa's somber face disappears to be replaced by Joe's somber expression of concern, back-shadowed with stark trees and dark clouds. Ow! Another painful squeal punctuates the silence. Rotating my head, I search for the source. Ms. Mizuikova's been reincarnated in the form of an irate squirrel! Her tiny woodland doppelgänger is stretched out to maximum height and its extremely bushy tail is flamboyantly twitching disapproval at me from few meters away.

"Do you recall your grandfather ever mentioning the name of this woman or any other female acquaintances?" Joe presses me.

Tatiana, you need to stop foolishly revealing details of your life. Don't let emotions for Joe and annoyance with your father cloud your judgment. "What are you planning to do to my grandpapa, Joe?"

"The real question, Anna, is what is your grandfather going to do to me?"

Yikes! That blasted squirrel chooses now to race past, chattering in its high-pitched chirp and practically jumping on my foot before scurrying up a tree. This menagerie of animal encounters, butterfly, hawk, squirrel, is very odd. They usually avoid me with a wide berth, just like people.

"Penny would be barking her head off right now and giving chase." Joe's face lightens as he speaks of her. "She treed a porcupine once, scared the heck out of me. I thought she'd end up covered in quills. Look, Anna." Joe directs my attention to the left. "Over by the tree line, there's a rafter of turkeys."

Turkeys, dogs, porcupines. Next, we'll be accosted by flying monkeys, lions, tigers and bears, oh my. *Not your circus, Tatiana, not your monkeys.* I'm starting to wobble. We bulls wobble, but we don't fall down. *Blin!* These meaningless Hannah expressions are going to be the death of me yet.

"Why are people and emotions so difficult to manage, Joe? How do you do it? How do you cope with all these chaotic feelings?" I'm starting to feel woozy. My unsteadiness increases, tilting me sideways. Clenching my fists, I pound my thighs with aggravation at my inability to control my life. Why am I like this? Why was my mama like this, swayed so strongly by emotions that they killed her and took her secrets with her?

Chapter 45

Joe, age six

I'm sulking in the corner of the classroom. Mrs. Summers is at her desk, pretending to be busy. With only the two of us here, it's sickeningly quiet. Finally, I hear my mother's footsteps. As soon as she comes in, she spots me and beckons me over. I run to grab her hand.

"I caught him taking money from my desk," Mrs. Summers accuses me.

"I didn't take it, Mom. I was looking for Tommy's library book. She took it away from him, and his dad's going to kill him if he doesn't return it. The book's due today."

Mrs. Summers has a momentary look of confusion. "I asked Tommy four times today to put that book away. The children need to know that I'm in charge."

"You don't know what his dad's like. He'll go crazy when he finds out," I yell at her.

"That's not the issue here," Mrs. Summers scolds me. "Where is the money you took from my desk?" she demands.

"My son doesn't lie, Mrs. Summers," my mother says. "Did you see Joe take the money?"

"No, but it's gone, and he—"

"We're done here." My mother cuts her off and we head out of the classroom.

"Every mother thinks their child's a little angel," Mrs. Summers mutters as we leave.

"I know you're not an angel, my little Einstein," she tells me as we walk down the hallway. "But you do have a big heart, always sticking up for the underdog. We should have named you Albert."

Joe
Sunday afternoon, October 14

I catch Anna as she falls into me and hold her close. Her whole body is shivering. "Have you had anything else to eat, Anna? You've got to take it easy and give your body time to recover."

"I need you," she beseeches me, raising her face to mine and pleading with a blend of desperation and yearning, exacerbating the volumes of her vulnerability that have already flooded into me. The vulnerability of caring, the vulnerability of connecting … the vulnerability of love. I can't leave her like this, raw to the world and its ugliness.

"Do you remember when you told me you felt like you were falling off a cliff, Anna?"

"No, and I'm too tired to try to remember."

"It was that rather cold day this month, and you were wearing a blue blouse with a tan knit sweater, sapphire earrings and—"

"Don't clog my head with superfluous information," she cuts me off.

"What I'm trying to tell you is that I won't let you fall off that cliff." *Don't be stubborn, angry or ornery, Anna, let me help you.*

"You can't prevent everything, Joe." She wearily pushes away from me, holding me at arm's length and trying to brace herself against the mixture of heartbreaking emotions ripping between us.

"Then I'll fall off with you," I solemnly state, reluctantly conceding that whether I want it or not, I'm bound to her and where she goes I'll be following, in some manner or other.

"I don't want you to jump off a cliff for me, Joe. I want you to explain your hidden agenda and help me control my emotions." She drills me with a can't-you-get-that-through-your-head look.

Did she get hit in the head during her kidnapping and get all the sense knocked out of her? I can't control emotions! Not mine and certainly not hers. "In case you haven't noticed, Anna, my temper frequently causes me to do things I shouldn't. I'm not the one you should be asking to mentor you in managing your emotions."

"But how do you manage them some of the time? How do you live with all these fractious feelings inside of you? Just give me any bit of advice, because right now I'm failing miserably against this hellacious onslaught." Her deep-brown eyes beg me for any type of support.

I doubt she'll believe me, but I tell her anyway. "I color-code them by type, then categorize them by date, location and subject." She looks at me like I'm telling her the world's flat. Is there any way to explain this to her? My memories of the miserable outcomes the few other times I divulged this to anyone are treading water in the background. What the hell, here goes nothing. "I remember things, Anna, every detail of every experience I've ever had. It's my curse and that's how I deal with it, by boxing things up into tiny compartments and storing them away." I bitterly lay out for her my dysfunctional, functional method of coping. Yea, I was right, she's having trouble believing me. I'm not sure if it's the remembering-everything part or the way I deal with it that she can't grasp.

"When I was little, it was hard for me to distinguish my memories from my dreams since both felt so real. I started getting night terrors, or at least that's what my mother called them, but they tormented me in the daytime too. After an especially brutal episode, I couldn't stop screaming, no matter what my mother said or did." Reliving this episode is grueling, but Anna needs to hear it. "My mother held me for a long time, trying everything she could think of to console me. Ultimately, she broke down and started weeping with me. When my screams at last subsided, she clenched me close to her heart and whispered in my ear, 'Think bright, happy thoughts, Joe, and don't let the gloomy ones take over.' Following her instructions literally, I began making my memories different colors and compartmentalizing them into different categories. It helps the majority of the time." But not with Lina. I can't keep those memories tucked away; they randomly pop out any time of the day or night.

Anna's face fills with disappointment. "I'm not capable of doing that, Joe."

She needs something more concrete to hold on to. "Have you heard of freediving, Anna? It's when you don't use any scuba gear and just hold your breath as you descend."

"A little," she hesitantly says, probably worried I'll be overwhelming her more extraneous data.

I make an effort to pinpoint my point, at least as best I can with all the competing issues battling for my attention. "In constant weight freediving, which means you're not using any weights or equipment except for fins, at about forty feet below the surface, gravity shifts. Instead of feeling the water pulling you up, it appears to be dragging you down. It's extremely important to be aware of when you reach the bottom of your dive, because if you go too far, you won't make it back. You and I are free divers of the mind, Anna, and if we allow ourselves to get too deeply immersed in our memories, or in your case your feelings, we exceed the point of safe return. It's a dangerous rabbit hole to go down."

Instead of the baffled look I expected from her, she's glaring at me with angry frustration. "I can't do that either, Joe. The circumstances of my life, feelings that I don't want, don't understand and can't control, keep shoving me into those rabbit holes." She suddenly grabs the sides of my head and deadlocks our gazes. "It's only when I'm with *you* that the emotions can't force me down there. I need you, Joe."

The agony in her eyes seems to reach back generations. Shit! It's an impossible choice set before me. Stay and help Anna, or go and help Lina. If I stay, I may never find Lina, dooming her to a life alone, battling to simply exist, but if I go, I might be condemning Anna to a lifetime of being pushed down rabbit holes ... and one of those times, she might not be able to find her way out. How can I possibly choose between these two?

"You're thinking about her now, aren't you?" Anna asks, letting go of me. "It's the exact same mixture of agonizing heartache every time you do. Is that why you insist you can never be my boyfriend? Because with your perfect memory, it's never going to go away."

I flee to my fail-safe subject with Anna, discussing the cosmos. "According to astronomers, the coldest known place in the universe is a massive gas cloud five thousand light-years from Earth. At -458°F, the Boomerang Nebula is colder than the background temperature of space. However, they are wrong."

"I'll say they're wrong. The coldest place in the universe is inside of me," she laments.

"No. It's inside a black box in a small city east of Vancouver where a niobium computer chip is chilled to -459.6°F and—"

"Freaking cold!" she explosively yells. "I'm so tired of all the cold in my life." Pulling the hood of the sweatshirt up over her head and tugging on the cords, she covers most of her face. Yet concealing her eyes doesn't prevent her excruciating torment from leaking through.

My internal sparring match intensifies. I need to stay and help Anna! I need to go and look for Lina! Fighting these contentious, mutually exclusive choices is brutal. *Facts, Joe, facts.* "It's called the Infinity Machine, Anna, a quantum computer in which qubits exist in separate realities at the same time. All the realities are true, but—"

"No more realities, Joe," she abjectly objects.

"The human insistence on a definitive, objective, observable world is a fallacy. Reality has infinite possibilities." I don't deviate from my safe strain of conversation.

"I don't need more realities, Joe, I can't deal with the one I have." In a last-ditch effort, Anna throws back her hood, wildly clutches at my face and kisses me, sending an overwhelming amalgamation of feelings rushing between us. Excitement, anxiety, exhilaration, alarm ... love.

Love! I quickly pull back and awkwardly caress her face. I can't love you, Anna, not in this string of existence. No matter how many times I've rehashed it in my mind, there's no timeline where that can work out for us.

Latching on to my shoulders, she violently shakes me. "I need you, Joe. Can't you understand that?"

A hundred thousand volts of Anna angst invade every inch of my being. I understand all too much and all too well. *I can't leave Anna to be buried alive in a rabbit hole, and by helping her, I may be sacrificing Lina's chance for a future. Damn you, Kluge. Damn your shitty secrets and the uncompromising promises you cornered me into making, condemning me to hell.* "Alright, Anna, I'll help you. I'll tell you whatever secrets you want to know."

Chapter 46

There is one true story,
but it has many facets,
seemingly in conflict.

When the same event is given contradictory interpretations by different individuals, it is called The Rashomon effect.

In quantum physics, the Rashomon effect occurs when the same event is given contradictory outcomes by different observations.

The double-slit experiment. A group of electrons is shot through a wall with two thin, parallel, vertical slits in it. The result is an interference pattern, as if two waves are colliding and creating ripples.

The second time, the electrons are shot through one by one. Astonishingly, the result is the same, an interference pattern.

The third time, the electrons are "tagged" before entering the slit, in order to observe where they pass through and where they go afterward. There is no interference pattern! The electrons act like individual particles and form two distinct groupings.

The last time, the electrons are tagged before they pass through the slits, then untagged right after they pass through. The result is an interference pattern.

CHAPTER 46

So ....
Is there really only one true story?
Or is there another one,
separate from our examination of it?

Tatiana
Sunday afternoon, October 14

Joe is going to tell me his secrets! My mind is doing forward somersaults, my heart is doing backward somersaults and my stomach is along for the 'tumbly' ride.

"You need to see a doctor, Anna, to make sure you're alright." Looping his good arm through mine, he tries leading me back to the sedan.

I firmly dig my feet into the ground. "Don't pretend I didn't hear what you just said."

"Not here and not now." He nods to the guards.

I completely forgot they were there! How much of our conversation did they overhear? They're too far away to have made out what we said when we were talking in hushed tones, but whatever I wildly blurted is another matter. I wish I had Joe's ability to file things away and recall later with total accuracy, then I'd know which is which and what I said. A light bulb goes off—I can ask Joe and have his perfect memory fill in the gaps. It'll be interesting to see how much he really does remember.

He continues insistently tugging me toward the car. "I don't need a doctor, Joe," I protest. "My father took my mama to numerous ones, and none of them were able to help her. I'm not going to be examined like a lab rat when I already know the outcome. You, on the other hand, definitely need a doctor."

He breaks out in a hearty chuckle. How is that amusing? He raises his injured hand and nods at it. "The other hand," he explains.

His slap-happy humor is infectious and I start snickering, which unexpectedly leads to fits of giggling and then onto laughing so hard that the "good type of tears"—an expression which until this moment I have never understood— stream down my cheeks, washing away my problems. Temporarily anyway. After the last of the guffaws fade, I'm out of breath and amazingly uplifted. When I turn toward the car and see the baffled expressions on the guards' faces, another burst of giggles erupts. I can picture what's going on in their minds; *the crazy bitch has finally lost it.*

"That certainly was a *handy* joke," I jest, a little tipsy from all this unaccustomed jocularity. Wow, that's funny too, jock-ularity. It was only last week that Hannah explained the meaning of the word "jock" to me. As much as I'm enjoying this comic relief, it's time to sober up. I coax Joe into the sedan, with the ultimate goal of getting him medical care. *Huh?* This bout of fifty-proof hilarity must have him somewhat soused, too, because he settles into the back seat without arguing.

The car pulls out onto the street and Joe leans forward, instructing the driver to take us to "the hideaway." What? When did we exit off the main road of trusting each other and "I'll tell you all my secrets" to end up right back where we started, in this backwater, backwards lane of a relationship? Why is Joe taking me to my father? Did I completely misread what just happened between us? Catastrophic anger is bubbling up. *Get ready for the battle of a lifetime, Joe.*

"Anna, the Hideaway is a restaurant and you need food," Joe rapidly clarifies, alarmed by my intense reaction.

Whew! Tremendous relief floods through me, but a minute later, my temper starts ratcheting up again. Joe did exactly what I've been complaining to him about in crystal-clear, flashing-neon detail, other people unilaterally making decisions for me. He decreed not only that I needed to eat, but also where to eat, without asking me!

"Sorry, Anna. How about we grab some food along the way, okay?" He contritely amends his single-handed decision.

I nod my approval. Sometimes this emotional link isn't such a terrible thing, and on occasion, it seems to be an extremely effective, efficient method of getting through to him. If only it could work that way on my father. *Wait another minute!* The driver's already speeding along per *Joe's* instructions, without so much as a sideways glance at me for confirmation. *Grrr.* Well, on the plus side, that probably means neither of the guards overheard much of our conversation and still believe Joe's working for my grandpapa.

The drive to the Hideaway is short. The restaurant's downtown storefront is a pleasing pastel-salmon color with a complementary awning over the entrance touting great food, wine, beer and coffee. As we pull up to the curb, my stomach lets out a loud, rumbling grumble and my eyes zero in on a little sidewalk chalkboard listing some specials. Joe tells the guards to get him two Carleton panini sandwiches and a large coffee with cream but no sugar, then turns to me for my order.

He's a quick learner. Granted, this is an inconsequential decision compared to life altering choices, but still, it's mine to make. Hmm, a *Carleton* sandwich, I wonder what that's got in it? Lots of underlying connotations there. Why not, I'll risk it. "Carleton sandwich, kale salad and a wild berry smoothie," I say. Even though I shoved down a ham sandwich back at the safe house, I'm still famished.

Joe and I remain seated in back while the guard on the passenger side goes into the restaurant to place our order. The driver and Joe remain quiet while waiting, and I'm almost asleep when the guard returns, his arms filled with multiple bags of deliciously smelling food and a takeout container holding four large drinks. There's a huge smile plastered on his face, obviously approving of this idea and including himself and the driver in the evolution.

The second he's seated inside the sedan, Joe immediately tells the driver to take us to the parking lot closest to my dorm. Before I can argue, he gives me an uncompromising stare. "A hospital is not the place for private conversations, Anna," he whispers.

I'm half-tempted to insist he get medical attention first, his hand is looking really nasty and swollen, but I don't risk it because I really want to hear everything Joe knows before he changes his mind about spilling his secrets. In fact, I want to have heard it all yesterday, no weeks ago, actually since the day I met him. So, when I open my mouth, all I do is wolf down the incredibly scrumptious food.

The guard added a pastry to all of our orders, and it's beyond lip-smacking tasty. Mine's raspberry with white chocolate, Joe's is peach and both the guard's and the driver's look to be some sort of berry. It appears Joe has won them over on numerous fronts, a united endorsement of his food stop and a fellow kinship in their dislike of Medusa. Unlike her security crew, these two men, who arrived with my father, are willing to bend the boundaries a little. They need nicknames, something to call them other than driver guard and passenger guard. Let's see, both are redheaded, wearing black pants, a black sports coat over a blue button-up shirt and a bow tie. Tweedledum and Tweedledee. No, that might tick them off if they overhear me, even if they don't know those are characters from *Alice in Wonderland.* I don't want to squirrel their cordial attitude toward Joe and me, so the passenger guard will be Tick, since he seems to jump right into action, and the other one will be Tock, since he's usually a 'tock' behind.

I'm still shoving the last bite of pastry in my mouth when the car pulls to a halt in the school parking lot. Joe and I gather up the remnants of our late lunch, which I fully intend to finish, and Tick and Tock exit with us.

"Wait here and enjoy the rest of your food," I direct them. "Since Joe is working for my grandfather, he's not a threat and I'll be safe with him."

There's almost sympathy in Tick's voice when he replies, "We're under very strict orders from your father to keep you in sight at all times until he has finished his final assessment and initiated all the new security measures to fit the current circumstances."

Did my father make him memorize that line?

"We could jump into your car, Anna, and have some privacy from the tag team during the ensuing chase," Joe half-jokingly, half-seriously suggests in English.

Tempting, but not worth the consequences. "Let's check out my room and see if Hannah's gone." In short order, Joe and I are alone, seated across from each other, with my Tick-Tock tag team stationed outside the entrances of the building, well out of earshot. It was an amazing feat of negotiation on my part.

"What do you want to know, Anna?" he asks, all sullen and serious, the lighthearted humor from earlier gone.

"Where did you learn that poem?" I cut to the chase.

He shifts uncomfortably. "No one can *ever* know what I tell you, especially not your father or grandfather." He waits for my acknowledgment.

What am I agreeing to? I wish I had some way to read the fine print. How will my father or grandpapa be affected by Joe's actions in the future? Will I be able to prevent those consequences if I don't accept his terms? The weight of his requirement presses heavy on me. After a couple of minutes rolling over these things and testing how they settle in my gut, I nod my concurrence.

"I was six years old when my mother died," he begins, talking in a hushed voice even though there's no one else around to hear. "By that age, I had learned to manage my memories somewhat, but my emotions and actions were wildly unpredictable. My father had no idea how to deal with me, he didn't understand me."

The story of my life.

"Being left raising six kids on his own, he didn't have the time, patience or energy for me, but he tried. Stern talks, extra chores, groundings for days at a time and when I really drove him up the wall, a thrashing. But it didn't matter what he did, I kept getting into more trouble."

Similar themes as me, just different subplots. Not wanting to interrupt his monologue, I keep my many questions to myself for now, greedily soaking in every detail of his life that he's willing to share.

"When I was in fourth grade, I did something that shoved my father beyond his limits, and he had a long talk with the minister. They agreed the best course of action for me was to go every day after school and help an elderly German gentleman who had recently moved into town. Thus began my indentured service to Mr. Kluge."

Klyuch? "Klyuch means 'clue' in Russian. That's a strange coincidence." *Rats!* I'm supposed to be listening, not interrupting. *Please, don't get sidetracked, Joe.*

"After all these years, I never thought about that correlation," he muses.

"What did you do that was so bad?" I guide him back on topic, curious to find out what he did to infuriate his father.

"Every rotten thing I could think of to get Kluge to kick me out so I wouldn't have to go there anymore." There's a half-amused grin on his face.

Plenty of juicy details there, I'm sure. However, I'll save that sidebar for later. "No, I mean what did you do that got your father so mad?"

Joe lets out a disparaging sigh. "I decimated our chicken's egg production."

"What?" I almost spit out the last bite of my sandwich as a few chuckles work their way up to cluck out of my mouth. "Why did you do that?"

"My older sister, Marie, was always angry with me because she got saddled with the responsibility of watching over me and consequently ladened with some of the blame when I inevitably got in trouble. Shortly after her fourteenth birthday, she got herself a paid 'nanny' job, as she referred to it, and not only succeeded in convincing my father that I didn't need babysitting anymore, but also successfully argued that she wouldn't have as much time around the farm so I should be helping with more of the responsibilities—like taking care of the chickens, which was one of her duties. I got stuck doing all the chicken chores on top of all my other chores and *none* of my responsibilities got passed down onto younger siblings."

"You didn't slaughter all those helpless creatures, did you?" I ask in half horror, perched on the edge of my seat. Unlike seagulls, I have a soft spot for chickens.

"No, at least not intentionally, and only a few of them were collateral damage." He pauses to take a swig of his coffee. "When no one was watching, I'd bring the dog into the henhouse with me to 'help' round up the hens. In a productive way for me and a counterproductive way for my father, it made the eggs easier to collect. Also, I'd sneak out at night and let the barn cats into the coop for added 'protection' against raccoons. The extra stress in the chickens' lives did the trick and their egg-laying capabilities were dramatically reduced, decreasing the amount of work I needed to do. As soon as my father noticed the significant drop in the number of eggs, he kept a closer eye on me, and more creativity was needed on my part. I began leaving some of the eggs in the coop to make the chickens start brooding, and in addition to that, unscrewed the light bulbs so they wouldn't turn on at night."

Noticing my confusion about the light bulb bit, Joe expounds on it. "In winter months, at least in Northern Minnesota, chickens don't lay eggs as much because of the shorter daylight hours. The light bulbs aid in preventing that. Anyway, my father eventually figured out everything I was doing, and boy, did he get royally bent out of shape."

I'm enjoying this intimate peek into the inner workings of Joe, but I want to get back to the subject of the German guy. "What rotten things did you do to Klyuch?" I ask.

A mischievous grin returns to his face again. "I'd *forget* to close the doors when it was cold out, I'd *accidentally* spill sticky things in the kitchen or on his desk, and I very purposefully cut holes in his socks and the bottoms of his garbage bags."

A vision of a young Joe cutting holes in socks brings a smirk to my face too. "That might be the source of our attraction. We both had a youthful fetish for cutting holes in socks," I tease. "My reasons for doing so were purely altruistic, though. I thought I needed to keep my cat's neck warm in the winter, and my socks were too small to do that, so I had to use my father's."

Joe glances at his watch, restless about something. "Where did you learn that poem?" I ask again before he tries to slip away. Learning about his childhood is insightful and amusing, but I want an answer to this pressing question.

"I'm getting there, Anna. After a lot of attempts, I finally got Kluge really upset. Grabbing me by the collar, he exploded and screamed, 'Yeban'ko maloletnee, sookin syn, Svoloch.' "

"He's Russian!" I blurt. The intrigue level of this tale has reached electrifying.

"Quiet," Joe admonishes me and glances at the closed door. Tiptoeing over, he whips it open, steps into the hallway and searches both ways to ensure no one is nearby. Satisfied that no one overheard, he comes in, locks the door and returns to me. "Yes, he was Russian, and he made me swear on my mother's grave that I would never tell anyone."

Footsteps sound on the stairs, then in the hallway. We hold still as they approach, pass by and fade away. "That's how you learned to speak Russian without your family ever knowing about it," I whisper. Things are starting to make more sense. Now who is this Klyuch and how did he gain such power over Joe? It's killing him inside to break his vow to this man. "Is he the one who taught you to fight?" I ask.

"Yes. 'Never underestimate your enemies, Joseph.' Those were Kluge's constant words of wisdom to me, and he followed them to the extreme, assuming everyone was his enemy. He was a bitter, paranoid old geezer who coerced promises from me that have haunted me since the day I made them," Joe berates Kluge, then curses under his breath.

"Who did you kill and why?" I spurt out, yearning to get all my crucial questions answered before we get interrupted again.

"He was a gang member that broke into Kluge's house, looking for cash or something valuable to hock. The old man was lying on the floor when I entered that day, blood pouring from his head. His attacker grabbed me from behind and slammed me against the wall. I managed to get ahold of my switchblade and stabbed him."

Another bulb goes on over my head. Two lighting up in one day, that's a record around Joe. "Lina was afraid you would kill her dad," I murmur, suddenly comprehending why she left him so abruptly and why it affected Joe so deeply.

"I couldn't let that asshole of a father keep beating her and do nothing, and neither could Kluge, but I wouldn't have killed her father," Joe defends himself, fiercely wanting me to believe him.

"Klyuch knew Lina?" Momentarily I'm stunned, as more of the intricately hidden web of connections spin together.

"It was the only reason he wanted me around, because I was friends with her." Joe shakes his head in frustration. "I never bought into his delusions of persecution. I always thought it was Lina's father I had to protect her from, but after all we've been through lately, I'm not so sure anymore. Maybe he wasn't as paranoid as I thought."

"Wait a minute, back up. Who exactly is Klyuch, and how was he connected to Lina?" I tilt forward, straining to catch every syllable so I won't miss anything, no matter how inconsequential. I may need every single piece of information to figure this all out.

"He was a sadistic, old bastard," Joe sharply snaps.

Parallel feelings of bitterness from all the deceptions, subterfuge and duplicity in my life rise up, and I heartfully commiserate with him. Moving my chair right next to Joe's, I reach over to touch his hand that's tightly grabbing his knee, wanting to dilute some of his disconcerted tension.

Joe relaxes his grip slightly and continues. "After the break-in, which Kluge insisted was connected to the Russian regime trying to track him down, he decided to burden me with the real details of his life. When Kluge was a little boy, Stalin's secret police arrested, tortured and executed his father. His mother, older brother and he were sent to prisons, and Kluge was the only one that survived. Eventually, he ended up at an orphanage, did extremely well in his studies, and after Stalin's death, entered Moscow State University. Graduating with a degree in engineering, he went to work at the Mir diamond mine."

"The Adámas!" I incredulously gasp, astonished at how this intertwining of our lives is unfolding.

"Who?" Joe's undivided attention fully rivets on me. *Blin!* Why must I do that around him? I never leak words any other time. "Who's the Adámas?" he repeats.

"It was a name I overheard late one night when everyone thought

I was asleep," I reluctantly reveal. "The following day I asked my grandpapa about it, and he became extremely alarmed, saying I could never mention that name to anyone or all our lives would be destroyed." A pang of guilty fear causes me to cringe, and I hold back the rest of the information, worrying that it still could endanger our lives. *T'fu.* I'm doing exactly what I've been accusing Joe of doing, withholding secrets. Additional self-reproach hits me as I *feel* exactly what I've been coercing him into doing, breaking a sacred promise.

"You know more than what you're telling me, Anna. I can feel it. Who is the Adámas?"

Chyort! I don't want to betray Grandpapa and my family. *Bzzzt ... bzzzt ... bzzzt.* My cell phone's vibrating, a fortuitous way out of not answering his question.

"What's taking you so long, Tatiana?" My father barks out so loudly I have to move the phone away from my ear. "I thought you only had to ask Carleton if he was going to remain here and continue working for your grandfather?" He makes this sound like a simple "yes or no" type of question. It's not, at least when Joe's involved.

"We stopped to get food and we're in my room now. The two guards are with us." Or at least sort of with us. "You don't have to worry, Father. I'm fine." I attempt to placate him and hold him at bay.

"Does that mean you finally made a decision? You're staying at Carleton?" he questions. Joe's eyebrows rise at that.

Is it what I mean? "Yes, Father, I'm remaining at Carleton."

"And you're not going to change your mind again?" he booms in an exasperated voice.

"Not anytime soon."

"Then I'll finalize the new procedures for you. Don't do anything rash, and stay put until I'm done," he annoyingly dictates, then ends the call.

"Who's the Adámas, Anna?" Joe immediately peppers me. "I'm

breaking promises I swore on my mother's grave by sharing my secrets with you. I deserve to know." He squeezes me between two twisted, incompatible loyalties.

What do I do? As I look at Joe, realizing the full extent of trust he's placing in me, I opt to take the plunge and confide in him. "The Adámas was a man who disappeared without a trace after stealing a fortune's worth of diamonds from the Mir mine. The Soviet government couldn't track him down, therefore they never acknowledged it happened, and my grandpapa ..." My voice trails off. It's hard to go against his wishes and my pledges to him.

"Your grandfather was directly involved in the investigation, wasn't he?" Joe readily puts two and two together, reaching this conclusion.

"I'm not sure," I hedge, truly not knowing why my grandpapa had some of the diamonds or how he was connected to the Adámas. I never asked him about it again and he never mentioned it. Deftly, I attempt to shift the conversation. Joe's been a good instructor in the art of how to avoid uncomfortable subjects, even though his tutoring on this skill has been inadvertent. "Klyuch is the Adámas, isn't he?"

"Adámas, what an interesting pseudonym," he scoffs. "Adámas is an ancient Greek word from which the word 'diamond' is derived, meaning unalterable, unbreakable or untamed."

That is an unusual choice of word to label him. I never would have looked into its etymology though. Being around a walking encyclopedia occasionally has its perks. "Why did Klyuch steal the diamonds?"

"He was married and had a child. His wife was extremely outspoken against the Russian regime, and that was creating problems for them. Kluge was in the process of working out an escape plan when she was diagnosed with a terminal disease. She made him promise that after her death, he would take their daughter somewhere she could live free from the long arm of the Soviet regime. He stole the diamonds, gathered up the baby, worked his way across the continent, and eventually ended up in the US." Joe succinctly concludes the biography, but it's not the final page of the final chapter. He lived to be an old man.

"Is he still alive? What happened to his daughter?"

"He's dead and she's dead." Distractedly, Joe glances at his watch again.

It can't just end there, too much is still unexplained. "Did he ever remarry or have any other children?"

"No."

"Then why is it so important for you keep his secrets?" That unique Joe mixture of guilt, loneliness and self-condemnation, rushes into me. "Lina, she's related to the old man!" I exclaim.

"She's his granddaughter. I promised to keep her safe ... and I've categorically failed." Joe hangs his head, mentally flogging himself.

"It's not your fault," I awkwardly state.

"Yes, it is. I should've told her all these damn secrets a long time ago." Joe's voice is starting to break. "Damn Kluge! Damn him, damn him!" Swapping to physical flogging, he slams his good fist into the chair with a hurtful power.

I jolt up as I vicariously feel his pain. "Does Lina know anything about the diamonds or Klyuch?" I'm ashamed to keep asking him about this painful time in his life, but there are still important questions I need answered, and waiting until later won't make it any easier.

"No, she doesn't know anything about her past, not even that Kluge was her grandfather. Her birth mother died shortly after she was born, and her father hated Kluge so much that he wouldn't allow him to see her. At least that's what Kluge claimed, and he never tested the theory. The old man kept a low profile and stayed inside most of the time." Joe shakes his head at the lunacy of it all.

"And Lina's father never told her about Kluge?"

"Her father doesn't know much about Kluge's life, because Lina's mother—her father's first wife—didn't know anything of Kluge's past. Can you believe that? Kluge never told his own daughter who she really was. The old bastard firmly believed that if she had carried the weight of all those secrets, she'd never have had the possibility of a free, happy life. So he dragged them along solo until unloading them on me."

This crisscrossing of secrets keeps getting more and more tangled. Why do people make such irrational decisions based on emotions? "But that diamond heist happened an ocean away and over fifty years ago and no one ever managed to track him down. Why didn't he at least tell Lina he was her grandfather?"

"Kluge insisted it was way too dangerous, claiming they'd never quit hunting for him and the missing diamonds, and they'd stop at nothing to get them, including torturing his granddaughter." Joe gets a far-off look in his eyes. "I almost told her once."

"What stopped you?"

"What else? Kluge. He vehemently contended I'd be signing Lina's death sentence. Staring me straight in the eyes, he asked me point blank if I was willing to risk her life to prove I was right, that no one was chasing him anymore. What choice did that leave me? I had to keep his damn secrets." Joe drop-kicks a pink throw pillow that Hannah put on my bed yesterday, and it lands then slides under her bed. He instantly retrieves it and places the pillow back where it was while trying to hide the fact that he's looking at his watch yet again. "I need to get going, Anna. I have something I have to do."

"Now?" I jump up out of my chair. "What could you possibly need to do *right now* on a Sunday afternoon that's more important than finishing telling me how all of this ties into my life. I want to know why both you and my grandpapa know that poem, why my grandpapa left off the last line and why it's so significant to you. What's the connection with that song you sang to me? And what did Klyuch do with the diamonds?" I hope he had better outcomes with those cursed Silver Bears than I did.

Joe reacts like he was bitten by a rattlesnake when I mention diamonds. He probably knows where they are!

"You're sure there's nothing more you care to know about, Anna? Like if I have any gang tattoos, drink blood or can transform into a werewolf under full moons?" He quips at my extended list of questions.

"I'm sure there's more, but those will suffice for now." I don't think I could mentally process any more at the moment. I need time to digest everything he's already told me.

"I don't have time to tell you all that right now, Anna, the urgent care clinic closes soon. I'll explain when I get back." He quickly beats feet for the door.

"Get back from where?" I yell at him. I know he's not headed to see a doctor. I've used this same word trick many a time with my father, listing factually true sentences and placing them in a specific order to create a conclusion that isn't accurate. It rarely worked. Also, the emotional feed between us is sending a signal that doesn't jive; Joe wouldn't be feeling so ashamed if the urgent care clinic is his real destination. *T'fu.* Must we always be teetering on the rim of an abyss, constantly watching our footsteps to avoid the one misstep that will send us tumbling over the edge? I have the sick feeling it's inevitable, one of us will eventually have to fall into that pit of hidden deceptions in order for the other to survive.

Joe hesitates as he's unlocking the door and turns back to me. "I meant it when I said I'd never let you fall to your death, Anna. I'll give my life to prevent it."

I don't know whether to be grateful or hateful at how this bond communicates for us, without the use of words. "We're finally trusting each other, Joe. Why can't you tell me where you're going?"

His mood swirls about like tumbleweeds in the desert. When the dust settles, he comes over and stands near me. "Kluge was diagnosed with terminal liver disease, rethought his decision about never revealing anything to Lina and concocted a half-assed idea to share a part of her heritage with her. He wrote down that poem, then mailed it to her without any return address, any signature or any explanation. It totally freaked her out."

He's avoiding my most recent question about where he's going, but that's okay for now. This is far too compelling to risk skipping over. This information is key to finding out why both Klyuch and my grandpapa hold that poem so close to their hearts. "Which part of her heritage? Who wrote that poem?"

Joe fidgets uneasily, reluctant to continue. "Kluge's wife wrote the poem for him shortly before she died. It was the last thing she gave him."

I determinedly follow the breadcrumbs he's dropped. It takes a while before it hits me. "My grandpapa must have known Klyuch's wife!"

"It certainly appears that way," he agrees.

"How, when, where?" I start pacing the room, zooming round and round Joe. He keeps looking at his watch. What the heck? *Don't get distracted, Tatiana.* "Tell me everything you've found out about my grandfather's relationship to Klyuch's wife."

"Are you absolutely sure you want to hear my suppositions?" The heaviness in his voice is offsetting.

"Absolutely," I unequivocally answer.

A bout of intense silence follows, and I feel every torturous emotional swing within him as he fights to figure out what to tell me. The corner of his mouth starts twitching. Well, at least he's not spooling off scientific theories. I wait ... then I impatiently wait more ... then I very impatiently wait even longer. At least four minutes have pasted by now. Alright, time to use a Hannah phrase, "We're not digging tunnels to China here, Joe, I'm absolutely sure I want to know," I sternly repeat.

His lips form a thin, tight line. "O.K., Anna, I think there was a notorious reason why your grandfather didn't write down the last line of that poem, 'we are immortal.' It was because his deepest, darkest secret would be uncovered if anyone ever made the connection that Kluge's wife gave it to him."

I wait ... then I impatiently wait more. Adopting another Hannah mechanism, I stomp my foot while poking my index finger at his chest. "I don't want to laboriously ferret out this riddle myself, I want to know immediately. Just tell me."

Joe whispers so softly I can barely hear him. "I don't think Kluge is biologically related to Lina ... I think your grandfather is."

My grandpapa had a child with another woman? I stand stock-still in shock.

"That's the only way all the pieces fit together, Anna. It's why your grandfather knows the exact words to that poem. It's why Kluge and the baby were able to flee all the way from Siberia to America undetected, because your grandfather side-tracked the KGB's investigation. It's why Kluge always felt like he was being tracked, because he was, your grandfather was keeping tabs on *his* daughter. And it's why, after all these years, no one ever found Kluge. Your grandfather wasn't really looking for him, he was only watching over his own daughter and then his granddaughter."

Valiantly, I use Joe's technique of sorting all the fragments into differently colored, organized compartments within my head, but the process is beyond me, and the whole mess quickly muddles into a disorganized array of errant piles. "I ... I ..." Reaching for support,

I unwittingly grab his injured hand, causing him to cringe in pain. Glancing down, I'm alarmed at how hideously swollen it is. "You have to get seen for that right away. I'm taking you over to the emergency room."

"No, I can get there myself, Anna." He callously pushes aside my offer of aid.

But when and how long from now will you get there, Joe? And what will you be doing beforehand? He needs medical attention more than I need an answer to where it is he wants to get to in such a hurry. "Just do whatever it is you're wanting to do *after* you see a doctor," I stubbornly argue. "I have my car keys with me and it's parked close by."

"You make it sound so simple, Anna. My life doesn't work that way. It's complicated and difficult and confusing." He runs the fingers of his good hand arrantly through his wavy hair, making it attractively tousled.

Not what you should be focusing on, Tatiana. The matters at hand are far more pressing than sexy hair.

Joe abruptly heads for the door again, sending parting words over his shoulder with a sense of finality. "I won't be seeing you much anymore, Anna. Odds are, I'll be kicked out of Carleton and sent to prison when the bloodstains from this morning's murder come to light. So whether I get my hand looked at now, or not, really doesn't matter in the scheme of things."

What scheme of things? There's still *something* that matters to him quite a bit, and whatever that something is, he intends to take care of it pronto without me. Will these blasted secrets never end? "I highly doubt you'll get sent to prison, Joe. Now that my father thinks you're working for my grandpapa, and my grandpapa's going along with that assumption, you won't get framed for that murder. They'll arrange for the evidence to point elsewhere."

"But I'll get kicked off the football team for fighting again, because I sure as hell can't tell the truth and explain the real reason for my black eye and busted lip. Saying I was abducted by your father and taken downtown to witness a killing wouldn't have a good outcome. And football is the part of my world that keeps me going, Anna. I could care less if I get kicked out of college, I'm only attending Carleton for a better chance at playing in the NFL."

"I'm taking you to the emergency room." I assertively insert myself between him and the door. "We can convince them you got a concussion when you were smashed into at the game and got your hand hurt. That way, you'll be required to stay in your room and rest, no going to classes or practices or anywhere else until the bruises go away. No one will ever know."

"More secrets and deceptions? That's your solution?"

"I just want you help you, Joe." Suddenly, all of the fatigue catches up with me, and I'm so tired I can barely move. "You claim you won't let me fall off that cliff, but how are you going to stop that if you're not here at school anymore? Isn't that just another deception?"

"Alright, I'll go get medical attention, Anna, if you will too."

Chapter 47

Joe, age sixteen

I've just finished reading an extremely lengthy biography about Albert Einstein for the essay I need to write, and I'm sitting here stunned, my high opinion of him destroyed. The revelations of his personal life are as fatal as the most venomous snakebite to me. Why did my mom make him sound like such a hero? Einstein cheated on both of his wives, and he had a daughter out of wedlock whom he never publicly acknowledged, probably never even laid eyes on. The man I idolized, who has been my silent mentor all these years, was a total ass.

Next day at lunch, I still feel like my world's been crushed. Lina instantly notices when she sits next to me in the school cafeteria. "You look like Penny ate your homework," she teases, hoping to cheer me up.

"It wasn't Penny that ate my homework, it was Albert Einstein," I grimly say. "After reading his biography, I can't stomach writing an essay about him. Einstein was an adulterer who abandoned his first child."

"He did a lot of good things though." Lina scoots closer to me. "In 1933, he wrote letters to countries asking them to take in unemployed German-Jewish scientists, and over the course of the war, he saved more than one thousand lives. Also, he supported the NAACP and publicly denounced racism, saying it was America's worst disease, and that was during a time when almost every other prominent public person remained silent."

"Since when did you become an Einstein expert?" I ask.

Lina shrugs. "You're always quoting him, so I've been reading up on him."

"Well, I won't be quoting him anymore."

She sketches a little heart on the back of my hand with her fingernail. "There are no perfect people, Joe. Martin Luther King Jr. and John F. Kennedy had affairs too, however, just like Einstein, they had a powerful, positive influence on our nation, inspiring us to change the way we view the world."

Joe
Sunday evening, October 14 to Monday, October 15

Three hours later, we finally return from the emergency room. Anna was slightly dehydrated, but other than that, no serious issues. I'm not so lucky. I'll definitely need surgery on my hand when the swelling goes down.

Anna starts vehemently pointing with both her hands to the oversized, and to put it politely, unflattering Medusa outfit she hasn't changed out of yet. "I'm dressed like a wallflower that was relegated to the compost pile weeks ago. Look at me, Joe, no one will come near me with a fifty-meter pole." Her protests, implying that I don't need to walk her to her dorm because no one would take a second look at her in this condition, aren't working. "It's important for you to go straight to your room and rest to pull off this concussion charade."

That argument isn't going to succeed either, Anna. There's nothing she can say that would convince me that walking alone at night across campus is a perfectly fine thing for her to be doing. In addition, this is my last bit of freedom for the foreseeable future since I'll be trapped in my tiny cell of a dorm room "recovering." And, there's still somewhere I need to go, downtown to follow up on the lead about Lina. I considered telling Anna, but it's too risky while her father and grandfather are continuing to have her closely monitored.

"You're going straight back to your room after we reach my dorm then, right Joe?" Her question sounds remarkably like an order. I give her a noncommittal nod. It doesn't fool her at all, however, she realizes there's nothing more she can do.

I search for something upbeat to say, not wanting to leave her with this air of distrust as her last memory of me. There's a remote possibility I might not see her again if my plan to find Lina goes awry. "From Curiosity, came the song heard round two worlds."

Her eyes narrow, in a good way, and her mouth quirks into its pondering expression. I can tell I've peaked her curiosity.

"Need another hint? Reach for the Stars, Anna, sang will.i.am. Although, he didn't actually sing your name. I added that in."

"I think I could have figured that out on my own, Joe. I'm not that gullible."

Yet she is curious, and she's stumped and doesn't want to admit there's something about Mars she doesn't know. "Will.i.am composed that song for NASA, and it was uploaded into Curiosity. Having reached the stars and touched down in Mars' Gale Crater the evening of August 5, 2012, PDT (morning of August 6 EDT), after an eight-month, 352-million-mile journey, Curiosity broadcast the first song in history from another planet and the second to be broadcast into deep space.

"What song was the first?" she asks.

"Across the Universe" by the Beatles. NASA beamed it to deep space in 2008 from its giant antenna in Madrid, Spain to celebrate the 50th anniversary of NASA and their 45th anniversary of the Deep Space Network. The song continues on its way *Across the Universe* to Polaris, the North Star, and won't reach its final destination until 2439 due to Polaris being 2.5 quadrillion miles away.

"Thanks, Joe." Anna gives me a quick kiss on the cheek, then disappears inside her dorm.

At least that ended on a positive note. I stroll back in the general direction of my room, breathing in the cool air and attempting to sort through my current collection of colossal screw-ups. I broke my promise to Kluge, telling Anna almost all of his secrets. How much danger have I put Lina in by doing that? How much danger have I put Anna in by doing that? From my recent interactions with her, it's clear she's not privy to all of her family's hidden, heinous skeletons. But neither is Lina! I've got to get downtown and search for her immediately ... yet there's a nagging concern that's holding me back. If I get caught faking a concussion, my temporary get-out-of-jail-free card the hospital granted me will get nixed. Then all kinds of questions will rain down on me, most of which I won't be able to answer unless I throw Anna under the bus, something I'm obviously incapable of doing. So, I've run full circle back to where I started. Will it be Lina and Anna? After another mental lap of second-guessing, I ultimately decide it's too late to accomplish anything tonight and return to my room.

"Where have you been, and why don't you ever answer your texts?" Lance angrily accosts me as I try to sneak in. "What are you thinking, running off half-cocked in the middle of the night and then ghosting me for an entire day? Oh, yeah, you don't like "thinking."

So much for a stealth reentry. "I was at the hospital, Lance, I had to get more X-rays of my hand." I exhibit my completely immobilized lower arm. "By the way, I was diagnosed with a concussion, therefore, I need to be on total brain rest to allow time for healing."

He gapes at my black eye and busted lip. "What were you doing that got you a concussion, Joe?"

It's best if you don't know, Lance. I don't want your future ruined too.

"I have a doctor's note for you to give to the coach." I place the paper on his desk. "I'd do it myself, but stage one in the recovery process of concussion protocol is total brain rest. No physical activity, no schoolwork, no looking at any electronic screens, no cell phone usage and no going to class or practice." I recite verbatim the instructions written on the paper.

"I'm tired of your games, Joe. Unless you tell me what's going on, I'm not chop-blocking to cover for you anymore," he inflexibly lays down his conditions.

"Anna's back," I quietly state.

"Back from what?" Lance demands. "Where was she? What happened to her?"

I can't stay in the pocket and not get sacked with all these questions coming at me, questions that I should avoid or fabricate answers for, so I scramble and attempt a Hail Mary pass. "She was kidnapped."

"Yeah right, and you, the shining Carleton knight in armor, single-handedly rescued her." He's not buying it.

"Her grandfather has a lot of money." I lie down on my bed and close my eyes. "I need to rest, Lance."

"Stop screwing with me, Joe, I'm done shoveling your bullshit. If Anna was kidnapped and her grandfather has lots of money, then she didn't need your help. What the hell have you been up to?"

Lance must be really upset, because he rarely swears. "Stage one of the concussion—"

"Go to hell!" He opens the door to our room, intending to storm out, then realizes how late it is and purposefully slams it shut as loud as he can. "And stay there!"

I feel like I've already been relegated there, Lance.

The next morning he's still fuming, so I feign sleep to avoid another confrontation. He does, however, grab the doctor's note I left on his desk.

The second he's gone, I get up and snag my wallet. I can't *not* search for Lina; the prospect of lying around here trapped in this little room for days or weeks on end with her out there all alone, facing who knows what because of me, will be hell on Earth. *You'd get your wish, Lance.* No matter the consequences, I have to go check out my only lead. As I'm heading out the door, I literally crash into Anna.

"What are you doing?" she demands in an adamant whisper. "You can't pull off this concussion diagnosis unless you stay in your room."

"I was opening the door to greet you," I innocently answer.

Taking off her scarf, she flits it in my face a few times. "I'm waving the red bullshit flag on your field of play and bringing in an umpire to contest it."

You're mixing up a few different sports there, Anna, and your scarf is deep burgundy, not bright red. Out of the blue, she jerks back in intense pain, shooting the sharp misery through both of us.

"Beckyyyyy!" Across the hall, a door slams open and a frantic student rushes out with her eyes pinned wide open in terror. Ping-ponging off the walls, she bolts through the hallway. "I can't believe this is happening! I've got to go to my sister now," she hysterically cries as she careens down the stairs. "There's nothing you can do," the guy chasing after her futilely tries reasoning. The commotion causes more students to come into the hallway.

'What's going on?' echoes from every direction. A few people pull out their cell phones to see what they can find out. "Oh my God, a major tsunami hit a nuclear power plant!" More voices ripple through the hallway. "They're evacuating the area ... Everything's destroyed! ... It's being called the worst nuclear accident since Chernobyl."

At the mention of Chernobyl, a panic-stricken Anna hurtles herself down the stairs, disappearing from sight. *Shit!* Another Catch-22 I'm stuck in. Obviously she's not all right, and I should go after her, but to have *any* hope of gaining *any* information from the one clue I have concerning Lina's whereabouts, I've got to head for Minneapolis. I can't put it off any longer, and with everyone distracted by the news of the disaster, it's the perfect opportunity to slip away unnoticed.

I surreptitiously proceed down the stairs. Anna will be fine, her father's still here and the bodyguards won't let her out of their sight. *Anna will be fine, Anna will be fine, Anna will be fine.* I keep chanting this to myself to tamp down the churning angst snaking through my gut.

When I'm only a few paces out the dorm, a nondescript, older gentleman falls into step beside me. A black fedora hat frames his steel-gray eyes, which are exuding weariness and have dark semicircles under them. I bet he's been up all night. *Join the club, mister.*

"We need to talk," he firmly states in Russian.

"No, we don't," I resolutely disagree, not caring who the hell this is. I don't need any more complications in my life.

"We need to talk," he persists as he rushes to keep up with my fast pace.

Who the heck sicced this guy on me? "I don't give a damn who you are, get lost and don't find your way back."

Haltingly, he starts singing in a low-pitched, gravelly voice. *"Where are you? ... I waited ... my entire life ... Do you hear? ... How to find ... my way to you?"* His breath is coming out in gasps as he strains to keep up with me. "Do you ... hear me ... Mr. Carleton?"

Unbelievable. The one person in the world that could sway my plans for the morning magically crossed the ocean to be here at this exact moment. I stop walking and resignedly turn to him. "Yes, I hear you, Mr. Alliluyev."

"We need somewhere to chat confidentially," he states. With a trained eye, he searches the area, then leads us toward Carleton's Skinner Memorial Chapel. Our footsteps resound loudly on the hard tile flooring as we enter the sanctuary of the English Gothic structure and move toward one of the wooden pews near the front. Anna's grandfather and I sit side by side in the looming space beneath the vaulted ceiling, silently readying ourselves.

Neither one of us knows quite how to begin. Without a clear direction, my encyclopedic mind takes over, slotting me into a certain archive selected by my surroundings and my hostility toward Mr. Alliluyev. Taking aim, I fire the first volley. "Goths were a Northern Germanic tribe that invaded the former territories of the Christian Roman Empire in the fourth century AD. Gothic buildings are larger than life, intended to invoke fear and awe." I tilt my head upward and he follows my gaze. "The pointed arch allows for larger windows to let in more light, but the word 'Gothic' now alludes to darkness. This seems ironic, doesn't it, Mr. Alliluyev?"

"What are you trying to tell me?" he harshly responds.

"Gothic architecture was designed to evoke feelings of insignificance and vulnerability, of being at the mercy of a higher power. So you choosing this chapel, which is quite an apropos environment for wrenching out deep, dark secrets, seems a bit fatalistic."

"We have far more important issues to discuss than architecture," he retorts as he takes off his hat and places it beside him. His thick, white hair is molded closely over his head.

I'm not finished with the Gothic topic yet. "Then I'll switch to Gothic fiction, perhaps you'll find that more to your liking. The plots have a persistent sense of impending, unidentified doom. Heroes are left to fend for themselves amid a mess of complex decisions, often in foreign countries. For the protagonists, estrangement from family is common, and often their family origins are hidden from them. Does that sound more relevant for you?"

"Are you quite satisfied now that you've had your say, Mr. Carleton?"

From years of experience, I know better than to answer rhetorical questions, but I want to piss this guy off. "No, I think you're a pompous, self-righteous ass who abandoned Syntyche Dobrow and the child you had with her to save your career."

Surprise momentarily shoots through his eyes before he recovers his composure. "You don't know what you're talking about."

"I know more than you think. I'm on a first name basis with *both* your granddaughters. Are you?"

"I'm the one who will be asking the questions here," he curtly instructs.

Good, I've got him on the defensive. I intend to make him pay for the havoc his selfish decisions have wreaked on Lina's and Anna's lives. "The granddaughter you decided to discard was beaten by her alcoholic father after her stepmother died. She ran away from him and has been living on the streets for over three years. All that happened because you're too much of a gutless prick to take responsibility for the consequences of your actions."

"You're even foolhardier than I thought." He glares at me.

"Thanks for the compliment, Mr. Alliluyev. I don't have any for you, neither did the man who believed he was Lina's grandfather. He was a mean, old bastard, but at least in his own warped way, he loved your daughter and your granddaughter." Mr. Alliluyev's strong front is weakening. It's time to go in for the kill. "Why did you help the Adámas escape Russia with the diamonds and allow him to whisk your daughter across the world?"

"You only have one facet of the story," he rebukes me in a measured tone, wanting to hide the shock that my mention of the Adámas caused him.

"Then tell me a different one," I challenge.

He pauses, intently studying his black leather gloves which are laying across his thigh. "You loved Ellie, didn't you?" he asks.

A ripple of fear courses down my back. He's using the past tense. "I love Lina," I assert. Lina's still alive. She has to be. *Please, let Lina still be alive.*

After straightening a wrinkle on his pants, he looks up at me. "You also love my granddaughter, Tatiana. So, where does that leave you, Mr. Carleton? Who will you choose?"

He's very cleverly twisted this back onto me. "My choice isn't as clear-cut as yours was, Mr. Alliluyev. I didn't get someone pregnant."

"Neither did I," he firmly denies.

I examine all his features, body movements and the inflection in his voice, using great attention to detail. I think he's telling the truth. "Then why did you put your career in jeopardy and allow Kluge to escape with the diamonds?"

Unbuttoning his overcoat, he removes it and drapes it over the back of the pew. There are spots of perspiration around the collar of his shirt and under his arms. Settling uncomfortably against the hard wood, he takes a moment to formulate his next move. "I did it because I never stopped loving Syntyche, and she never stopped loving me."

Not what I was expecting him to say. "So even after you were both married to different people, you kept the affair going?"

"It wasn't like that. We—"

"How do you know the exact words to the poem she wrote right before her death?" I cut him off, not wanting to hear his rationalizations. The door at the back of the chapel creaks open, and we both swivel our heads in that direction. A young woman peeks in, then immediately retreats upon spotting us.

When the door clicks closed, Mr. Alliluyeva employs a different tactic. "I'll explain some things to you, if you'll stop digging into this and give me your word you'll keep your mouth shut. You can't repeat what I say to anyone." He's grim yet assured, believing I'll agree to this.

"Hell no! I'm done swearing allegiance to sadistic, old men." I'm not foolish enough to make that mistake again. I'm wiser to the ways of the world than when Kluge manipulated me into doing his bidding.

"Do you want to know where Ellie is or not, Mr. Carleton?" He dangles this tantalizing carrot in front of me, except the tip of it has a poisonous ultimatum attached. Do I choose Lina or Anna? "What's your answer?" he coldly prods.

Here's my answer, Mr. Alliluyev. Getting up, I turn my back on him and start walking out of the chapel. I'm done with being trapped in zero-sum games.

"I'm sorry I had to do that," his hoarse rasp stretches out to me. "But I needed to find out where your loyalties are. For Tatiana's sake, please hear me out."

Was he really testing me that entire time? I spin around to inspect his face. Even though he's staring at me, he's looking past me, making it harder to determine the intent of his apology or even who he's apologizing to. Is it me, Lina, Anna, his wife, Syntyche Dobrow or all of us?

"Many years ago, I asked Syntyche to marry me," he confesses, his voice filled with heartache. "Unfortunately for me, Syntyche was strong-willed, high-spirited, outspoken and an extremely public voice of condemnation against the communist regime, an impossible fit for a rising KGB agent. She refused my proposal, telling me she loved me too much to watch my life get destroyed because of her." The agony of that decision still resonates deep in him.

Crap! That's the same damn justification Lina gave me. "Why did you risk keeping in touch with Syntyche if her daughter wasn't your daughter?"

"I never contacted her, I only kept track of her," he matter-of-factly explains, as if this were a perfectly normal thing to do. His affections are significantly more twisted than Kluge's.

"You loved her, so you spied on her behind her back. For what purpose? Certainly not to keep Syntyche, her daughter or her granddaughter safe, because you sure as hell failed at that."

"I monitored her from afar and never invaded her privacy," he continues ludicrously defending himself.

"Well you can stop monitoring, Lina doesn't need your kind of help. I'll text Anna that you've arrived on the scene." I turn to leave.

"Syntyche mailed that poem to me shortly before she died, and in it, she asked me to keep her daughter safe," he divulges, luring me back into listening. "It was the first and last time she communicated with me after denying my marriage proposal. How could I have refused the only thing she ever asked of me, Joseph?"

"Why did you switch to calling me Joseph?" Is this another one of his psychological ploys? Mr. Alliluyev's spent most of his adult life in the realm of subterfuge, and he's very skilled at it.

"I better understand the choices you've made, Joseph, and after meeting you ... I see a bit of myself in you," he earnestly replies.

What advantage is he hoping to gain?

Along the walls of the chapel are long rectangular banners, and on one of them is a vivid red rose surrounded by swirls of musical notes with the words *Led by the Spirit* at the top. Instantly, memories of Lina assail me, then Anna's presence bursts through. "Why did you sing that song at your wife's funeral, the same one Kluge sang on his deathbed?" My words rush out in a syncopated rhythm.

He hums a few of the notes before answering. "It was my favorite song, and it became one of Syntyche's favorites too." He softly sings a verse. *"For you, I'll surrender my entire life ... Do you hear?"* With full force, he slams his hand on the wooden pew, gaining my full attention. "Do you hear, Joseph? You need to let Ellie go, just as I had to let Syntyche go."

Why does everyone insist I need to let Lina go? "You don't know anything about me or Lina, and you have no idea what we need. You're just an old fool."

Astonishingly, after all the nasty things I've said to him this morning, this seems to cause him the deepest pain. "Do you honestly think I don't understand what you're going through, or the impossible choices you're being forced to make, Joseph?"

"Stop calling me that." I'm irritated, angry and unsure of what he's trying to do. Mr. Alliluyev is one of the rare people in this world I haven't been able to see through and spy out his underlying intentions. While his pain seems genuine, if he really cares about Lina, why has he never done anything for her?

"Psst, psst, psst. Maybe he's only after the diamonds," Kluge whispers to me from the grave as the distrust he drilled into me over the years splatters droplets of doubt. "I'm not making your mistake, Mr. Alliluyev, I'm not abandoning someone I love. Either tell me where Lina is, without me having to promise something in exchange, or I'm outta here."

"Letting go of someone you love is the hardest thing you can ever do, but it's not a mistake, Joseph." His tone is remorseful yet insistent. "For Ellie to see you, after what she's been through these past three years, will destroy her and utterly annihilate the love you both still have for each other. Is that what you want?"

The colorful light displays on the floor from the stained-glass windows disappear and the chapel darkens as the sun moves behind some clouds. "Exactly what has she been through these past three years?" I demand. "And if you've known what's been happening to her, why haven't you done anything?"

"I should have done more." Mr. Alliluyev heaves a sigh of remorse. "Ellie took off so unexpectedly and was so incredibly determined to disappear and remain hidden, that it was difficult to locate her. By the time I managed to find her, she was prostituting and addicted to heroin. I arranged for her to go into treatment, three different times, but she wouldn't stay. I have someone with her now full time, ensuring she's fed and safe."

"Where is she?" I demand through gritted teeth, barely able to contain my fury.

"I'm taking care of her, Joseph, and I'll make sure she kicks this heroin addiction. In addition, I've set up a trust fund for her so she'll never want for anything again."

His attempts to placate me aren't going to work. I'm not turning my back on Lina. "Where is she?"

"Rashly charging after her when she doesn't want to see you isn't a wise decision. Stop to consider the consequences of your foolhardy pursuit and the devastating affect it will have on Tatiana."

So that's his angle. He doesn't give a damn about Lina, only Tatiana. "Anna can survive without me," I push back. "She has plenty of people looking out for her, she's not all alone like Lina."

He stiffly stands and moves toward me. The deep creases around his eyes match his heavily wrinkled shirt, and his expression is a collage of pain, regret and determination. "There are different ways of loving people, Joseph. The love between you and Ellie is different than the love between you and Tatiana. I can and will help Ellie, but I can't give Tatiana what she requires, only you can do that. She needs you the same way my daughter needed Pavel, and he wasn't there for her. My daughter *died* because he wasn't there for her! Don't do that to Tatiana, please don't walk away from her, Joseph."

He's resorting to a combination of coercion and begging. I will not be duped into doing someone else's bidding again. "Tell me where

Lina is," I indignantly demand.

A beleaguered resignation fills his features and his whole demeanor changes. "700 Faith Way, Apartment 373, downtown Minneapolis. I'll notify them to expect you." Bowing over, he trudges back to the pew and collapses into it.

At last, I know where Lina is! With a mixed sense of excitement and a newfound determination, I stride toward the exit … until one last question makes me look back. "Did you love your wife, Mr. Alliluyev?"

"Very much," he mumbles, bent over in sorrow with his head in his hands.

"But not as much as Syntyche?"

Raising his head up and straightening his back, he firmly declares, "Every bit as much as Syntyche."

The old man had me going for a while, but this isn't real. He's had years to reform these memories into what he wants them to be. "You can't love two people that way at the same time, Mr. Alliluyev."

He spins toward me with a renewed vigilance. "No, Joseph, it's you who can't love two people that way at the same time, and it makes your choice infinitely more difficult than mine was."

"I wasn't given the opportunity to choose, Mr. Alliluyev. Kluge coerced me into promising on my mother's grave that I would take care of Lina and keep her safe. I'm not going to break that vow."

His eyes fire up with a fierce resolve. "I made that exact same promise long before you did. Ellie's my responsibility and I'm here now to take care of her. You've fulfilled your promise, Joseph, it's time to let go."

But I can't, especially now that I know where she is. I have to go to her.

He stands and faces me. "Don't turn away from Tatiana when she needs you most. She's waiting strong, but I don't know how much longer she can do that."

His poignant entreaties to help Tatiana are starting to have an effect. But I can't be there to support them both at the same time. How did Mr. Alliluyev do it? "What was the secret only you and Syntyche knew?" I ask, grasping at straws.

"What?" He's totally confused.

"The secret that only you and Syntyche knew, the last line from that poem."

"That our love is eternal," he instantly answers. Turning to the alter, he starts entreatingly singing. *"Where are you? I've waited ... Do you see? For you, I'll surrender my entire life ... Do you hear? How to find my way to you? ... For you, I'll light, I will love you eternally.* You sang those lyrics recently to Tatiana," he says in a gravely voice.

Not this time. No more scheming old men messing with my head. I rush outside to get away from him. Focusing inward, I search out a different old man, one who gives me a reassuring wink. *"Let the music flow through you and listen, Joseph. Listen, that is all."*

"That's not all, Mr. Einstein, it's more complicated than that."

"Her music has been speaking to you all your life, and even though she says no words, her song spreads out to the far reaches of this universe. When hope has disappeared, her spirit still sings and her inconceivably tiny strings of energy unite all who listen."

"Whose music and whose spirit? Who are you really talking about, Mr. Einstein?"

"Lina ... my Lina."

The ringing of my cell phone cuts in. Pulling it out, I glance down. The call's from Hannah.

"Joe!" Her screeching cry leaps through me. "There's something terribly wrong with Anna. You have to go after her."

"She's just upset about the nuclear accident," I brush her worries aside. Now is not a good time for more people confounding my decision making. I shouldn't have answered the phone.

"Don't patronize me. You're the one who insisted I call if I noticed anything strange going on, and there's something really, really out of whack with her. After disappearing for the better part of two days without so much as a how-do-you-do, she prances back in as if nothing happened wearing this God-awful outfit that looked as if she was in a sweat lodge losing fifty pounds. Then a little while ago, she raced past me on the green like a turpentined cat with its head cut off, appearing like death sitting on a tombstone hatching ghosts! She's never looked like that before."

"Was a bodyguard with her?" I curtly ask.

"Yes."

"Then she's fine." I'm trying to convince myself of this as much as I'm trying to calm Hannah down.

"That's it? You're not going to do anything? You're the cream of the crap and the crap of the cream, Joe. In case you can't figure it out on your own, that means you're a real jerk." She hangs up on me.

Anna can make it a few more hours, despite Mr. Alliluyev's and Hannah's assertions that she needs me immediately. She's managed to survive her entire life without me. Lina might not make it a few more hours, and she's been struggling through hell for three long years. This could be my last chance to help her. I can't pass it up.

With a strong sense of urgency, I head for the bus stop. *"Anna will be okay, Anna will be okay,"* I chant. With every step I take, the tangle in my stomach clenches and the pull to help Anna intensifies. This is ridiculous, I'm not sealing Anna's doom.

The bus to downtown pulls up to the curb and the doors open.

Chapter 48

Doubt thou the stars are fire,
Doubt that the sun doth move,
Doubt truth to be a liar,
But never doubt I love.
William Shakespeare

Tatiana,
Monday morning, October 15

After the news of the nuclear disaster, I race down the stairs in a panic with no idea of where I'm headed and no idea what's happening. I was only stopping by to tell Joe I'd bring him lunch around noon, when out of the blue, I was blindsided by a powerful attack of emotions, but they weren't mine or Joe's or my father's or anyone else's I know. Why would news of a nuclear disaster trigger an all-out personal meltdown?

Irrational anxiety worms its way inside me, spreading wildly wicked images that crawl through my skull. Crashing out of the building, I break into a run and mindlessly bolt across the campus.

"Anna! Anna!" Hannah's high-pitched squeal barely registers.

Increasing my speed, I frantically dart away from her, sprinting full bore until I can't breathe anymore. By the time my lungs finally give out, I'm deep in the throes of the arboretum. Glancing around, I spot a trail marker.

#42 Spring Creek: Spring Creek flows from southeast of Northfield through Lyman Lakes then into the Cannon River near the south end of the Lower Arboretum. As its name suggests, Spring Creek is fed by natural springs and never runs dry.

Never runs dry? The stream never stops? Wu wei is like the water of a stream. When a rock is dropped in, the water doesn't stop. *No! No! No!* No watery philosophies of wu wei running through me. Hastily spinning around, I end up slamming into one of my father's security personnel. I claw my way past like a feral cat and erratically hurtle away at maximum tilt until I'm again too winded to continue. Where of all places in this godforsaken eight hundred and eighty acres of wilderness do I end up? At another river!

Water flows around the rock and resumes its natural state. Flow with the moment. *No! No! No!* Recklessly, I charge deeper into the hinterland, racing as fast as I can while valiantly fighting this emotional genie that has been released from the bottle and is rearing its powerful spirit upon me with a vengeance. How can I trick it back into submission and lock it away for good?

This time, when my oxygen supply system forces a shutdown of maximum ambulatory speed, I make sure there are no rivers, no streams, no ponds, no lakes and no water of any type in sight. There's only a circle of large, knee-high rocks which I approach with part curiosity and part wariness. Surprisingly, my jitterbug nerves ease ever so slightly when I enter the formation.

"Deep within the center of your being, may you find peace. Gently, within the quiet of this grove, may I share peace. Powerfully, within the greater circle of humankind, may we radiate peace."

I instinctively round on the intruder. She's a middle-aged woman, with long auburn hair flowing past her shoulders, wearing a black and red macrame shawl laid over the top of a peach-colored robe that reaches all the way down to her ankles. She's standing a few feet from me with both her hands resting atop a waist-high, narrow, wooden stick balanced in front of her.

"That's a peace prayer," she softly explains, displaying a welcoming smile. She tentatively steps toward me, seemingly worried her movement will cause me to flee. "The prayer is a service of love, generated by giving and receiving."

"Vat eez dis place?" I ask, wondering where she came from. Although when I trounced in here, I wasn't in my usual state of full awareness, so she may very well have been here the whole time.

"It's a Druid stone circle for prayer and meditation." Sweeping her arms around in a fluid, inclusive manner, she broadens her smile. Reaching into a front pocket, she calmly pulls out a business-sized card and presents it to me.

"Prayer doz not vork for me," I sharply remark, ignoring her outstretched hand.

She stoops down, picks up a pebble and holds it out for me to see. "Prayer is like a stone dropped into a pond. The ripples spread until they reach the edges." She takes another step closer.

"E-nuff." I back away, covering my ears with my hands. No more water metaphors, no more wu wei.

"Stop, sense, see and feel the sacred cosmos within you. By touching the stars, you connect with your soul's immortality." The words continue rippling out of her, even as I'm running away from the stone circle. "Deep peace to you. Deep peace of the running wave to you. Deep peace of the shining stars to you." Her proselytizing phrases chase after me as I scramble with a sort of hysterical pandemonium back into the wilds, desperately trying to escape. Escape what? I feel so confused, so afraid and so alone. All alone.

Eventually, I end up charging into another clearing. Spotting

Stewsie Island in the distance, I gravitate toward it, barrel along, reach the bridge, cross over and approach the labyrinth. With a profoundly deep need, I navigate its twists and curves, desperately yearning for the enlightenment it promises. Upon reaching the center, I drop to my knees, close my eyes and wait, and wait and wait. I don't know if I can wait any longer, but I do. I wait stronger and I wait stronger, until ... I hear a sound!

Listening more closely, I realize it's me, crying. Tears are streaming down my face. I've reached the end. I have no more waiting left in me.

"I've waited ... Do you see?"

Where are these words coming from? Did that woman follow me? I try to look, but I can't see much with my blurry, bleary, tear-ridden vision.

"How to find my way to you? Where are you?"

"Where am I?" I sob to this voice that's probably only in my head.

"For you, I'll surrender my entire life ... Do you hear? For you, I'll light."

I'm conscious of a pressure on my shoulder. Exhausted and completely drained, I collapse, but I don't fall. A smooth, soft touch on my cheek gently absorbs my tears, and I look up to see Joe. His Siberia cobalt-blue eyes fill my entire vision.

Wrapping me close to him and holding me tight, he ardently and passionately pledges, "I'm here for you, Anna, and I will always, always be here for you."

Chapter 49

The best and most beautiful things in the world
cannot be seen or even touched—
they must be felt with the heart.
Helen Keller

Tatiana
Monday, October 15

Clinging to Joe, I suckle his strength, drinking in his healing elixir and allowing it to envelope my battered soul. Coddled in his comforting embrace, sheltered from the pains of this world, I begin to recover.

"I don't know what happened, Joe," I stammer through my lingering sobs. "I have no clue where these emotions are coming from, they're not linked to my mom, my dad, my grandpapa, you or anyone else I know. Foolishly, I thought I could run away from them, but I can't."

"We'll figure it out together, Anna. I'm here for you." His soothingly protective voice lovingly caresses me. However, there's a sliver-sized part of me that resists the wonderfully irresistible things he's telling me, because I know deep down they're not true. "You won't always be here for me, Joe. There will always be Lina."

"You and I have a special bond, Anna, one that defies space and time. I *will* always be here for you, I promise."

He promises that he'll be here for me! And his promises mean everything to him. Maybe, just maybe, he will. I want to believe it. Precariously, I tiptoe through the many fluctuating feelings inside of me. Pressing my hands to my heart, I gather them together and touch my lips to Joe's, fervently sharing my passions with him before pulling back to see his response.

Astoundingly, the clarity of his emotions that answer back are different than ever before. There's no conflict, no hesitation and no guilt. There's only ... love. Oh, it's so deliciously tempting, yet can I risk letting myself love him?

"No," the cold, sharp knife of reality cuts in; the perils are too great. Depending so heavily on Joe or anyone else in that way would cause me to become extremely vulnerable, and the odds of surviving the cataclysmic collapse of that type of relationship are so infinitesimally small, that for all practical purposes, they are nonexistent.

I try pushing Joe away, but he won't let me. Grabbing ahold of my hand, he bows his head, skims kisses of loyalty across my palm, then tenderly places our joined hands across his heart.

How can I make him understand without hurting him? He's been right all along, he can't be that special person in my life, no one can. Fate always has unexpected torments lying in wait for me, especially around anything connected with those damn diamonds, and Joe is intricately tied into them on so many levels.

My body begins to tremble, already feeling the gaping hole in my future, a future that requires learning to survive without him. *You have to let him go, Tatiana.* But not instantaneously, I can savor these moments. Staving off the inevitable a bit longer, I soak in all his love, hoping to store it away for those deep, dark times that are surely ahead.

Misinterpreting the fluctuating state of my emotions, Joe presses his lips to mine. I don't put up any resistance, instead I fervently join in. It's so incredibly glorious to feel loved! Aching for just a touch more, I slide my hands under his coat and shirt until I reach bare skin, then I melt into his warm body, letting my emotions take over. I want to feel his heartbeat next to mine every day, I want to look into his Siberian blue eyes every day and I want to hear him say "I love you" every day. For this short span of my existence, I allow myself to live all of that, tucking it away in my memory banks for all eternity.

"Ahem, ahem." An annoying coughing intrudes. "Ahem, hem, hem." The pestering noise won't quit. "Tatiana." I recognize Tick's awkward mumbling. "Ms. Mizuikova is asking where you are and what you're doing. She wants to know why you aren't in class. What should I tell her?"

Tell her to get a life of her own and get out of mine. How much longer is my father going to torture me this way? I reluctantly release Joe, grateful for the experience of a fairy-tale love I never imagined possible. "Tell her I'm heading back now."

Stretching my cramped muscles, I stiffly rise with some assistance from Joe. He interlaces our fingers, and I walk along beside him with a warm tingling stimulating every nerve in my body. As we reach the Bald Spot—what a nonsensical name for a campus green area—the Skinner Memorial Chapel comes into view, and the tingling sensation starts zinging out of control.

"Anna ..." Joe hesitantly begins, then stops talking and walking.

I knew it! That bloody ever-present guilt that dominates Joe's life is freaking back, and only minutes after he insisted he'd always be here for me. That didn't take long at all. Angrily, I whip my hand away. "Promises can't change what you feel, Joe, no matter how hard you want them to. You're always going to be consumed by that blasted guilt, and we both know it."

"You're wrong, Anna," he firmly denies.

The little devil on my shoulder is shouting out in a thunderously loud voice that I'm right. "After every kiss we share, the exact same gut-wrenching, poisoning self-reproach takes over your heart. You love Lina, and you always will." I should be glad for this; the permanent division should stop me from pining for the impossible, but it's not turning out like that at all. I'm devastated. Like a mortally wounded animal, I want to find a secluded spot to curl up in and lick my wounds.

"No, Anna. Every time I kiss you, the rest of the world disappears. Those brief breaks are the only times during the past three grueling years that I've been completely free from the regrets, the self-reproach and *all* thoughts of Lina. That's why I felt guilty afterwards."

"And what's different now, Joe?" I demand, not seeing the "guilty" distinction between what he said and what I said.

"What's different now is that I realize the love I have for Lina isn't the same as the love I have for you."

He loves me, Joe loves me! He's never said those words out loud to me before. *No, Tatiana. You don't have the strength to be rejected ... and you know you will be.* "This morning, when I went to your room and you smashed into me, where were you going?" He dodged my question then, will he do it again?

Joe nervously shifts his weight from foot to foot, standing there filled with apprehension, "It doesn't matter anymore, Anna."

"It matters to me. Where were you going?"

He sheepishly hangs his head in shame. "I was going to look for Lina, but—"

"Nothing has changed," I bluntly cut him off.

"Everything has changed, Anna. I'm no longer a prisoner to the oath I swore to Kluge." Cupping his hand around his mouth, he brashly yells at the top of his lungs, "I love you! I love you, Anna!"

I love you, Anna! His declaration of love, heard round the campus, skips into my heart and blooms with an exquisite ring. Excitement and hope race through me. Maybe? *No!* However true his love may be, it's not love that is meant for me.

"Did you find Lina?" I ask.

"In a way," he hedges with a confusing melancholy and an edge of misgiving mixed into his indeterminate answer. Placing his lips next to my ear, he whispers, "Your grandfather told me—"

"My grandpapa? You said you weren't working for him," I lash out, not caring if Tick overhears.

"I don't work for him, Anna," he says in a barely audibly whisper. "Your grandfather approached me this morning for a totally different reason."

What? I firmly shake my head no. "My grandpapa would not fly across the world to talk to you if you don't work for him or even know him. First off he hates flying, and secondly, if he did actually make that journey, he certainly wouldn't talk to you before coming to see me. I know my grandpapa better than that."

"Maybe you don't know him as well as you think you do," he counters. "He's quite a confounding, contradictory man."

Is my grandpapa really here on campus? I feel jealous and betrayed. "What's so outrageously important that the first thing my grandpapa had to do when he arrived here at Carleton was talk with you?"

Joe looks at me through shuttered eyes. "Your grandfather demanded rather forcefully that I not tell you, Anna. You should ask him that yourself."

"I'm asking you." Anxiety is restricting my vocal cords to such an extent that I don't recognize my own voice. "Why must your twisted promises always come first, Joe? Can't you see those misplaced loyalties are hurting more people than they're protecting?"

He sets his jaw in firm grimace. Thinking he's about to dig his heels in and refuse to share any information with me, I'm astounded when he softly confides, "Your grandfather explained his connection to Kluge's wife and his relationship to Lina."

Ancient skeletons that have been silently hanging within our family's proverbial, hidden closet start rattling about with an ominous, raucous warning. "What connections, Joe?" None of what he's saying makes any sense. Why would he tell that to him and not me?

He points the way toward the Skinner Memorial Chapel. "It's best you hear it directly from your grandfather. He was in there a little while ago, and I imagine he's still there." Touching my shoulder, Joe stares deep into my eyes. "If he won't tell you what you want to know, Anna, I will," he earnestly avows.

A weary moan escapes me as I brace myself for whatever is to come. With heavy steps, I trek toward the chapel, solo, and step inside. Hesitantly, I amble up the aisle of the sanctuary as legions of butterflies churn in my stomach and swarms of bees gather in my bonnet. In an odd sort of way, I'm starting to find some comfort in Hannah's animal idioms. They are marginally helping to limit the disruption in my emotional equilibrium. I have very little skill in that area and need every bit I can get.

My grandpapa is seated toward the front, on the left side. He's bowed over and still as stone. I continue forward, the only sound being the echo of my steps. The atmosphere is eerie and disconcerting, causing me to stop a few meters away from him. My rash decision to rush right in without thinking through what to say to him doesn't seem like a good idea now. Do I really want to know the secrets he's been keeping?

"I'm glad you came to see me." My grandpapa's sudden greeting startles me, resulting in my unconsciously jerking backwards. "Sit with me, Tatiana." He invitingly pats the space beside him.

I settle into the pew, wrestling with what to say. Hannah's cat has got my tongue and won't let it go. Or is it Schrödinger's cat? For in this instant, my grandpapa is both my nemesis and my idol, and it's not until I learn the truth, breaking open Pandora's box of emotions, that he becomes defined as one or the other. "Did you have an affair and a child with another woman besides Grandmama?" I blurt out, wildly ripping open that box.

There's a long pause, then he orders Tick, who is standing at the back of the chapel, to wait outside. After he exits, Grandpapa studies my face. "Joseph went right to you after he left here, didn't he?"

"That's a "yes" then. Who did you have the affair with?" I demand.

Grandpapa purses his lips and frowns with a somber thoughtfulness. "Syntyche was my Lina," he soulfully replies.

His Lina? What kind of an answer is that? "Why did you marry Grandmama if you loved someone else?"

"I love them both, very, very much." He looks hard into my eyes, just like he did so long ago before he fled to England, when he tried to explain things I wasn't yet capable of comprehending. "I was a young man filled with unrealistic dreams when I fell in love with Syntyche Dobrow. We had a magical year together, but when I asked for her hand in marriage, she refused me. Syntyche accepted what I couldn't at the time, that our very different lives would destroy us if we stayed together. We went our separate way, but I couldn't erase my love, and thus, I kept a distant eye on her to protect her."

"Protect her from what?"

"From the Russian regime, which she recklessly and antagonistically voiced her opposition to." A pained expression fills his face. "Alas, I couldn't shield her from the ravages of cancer. Right before Syntyche died, she wrote me a letter."

The chapel door cracks open, and Tick signals to my grandpapa.

"Get out and don't interrupt us again," Grandpapa authoritatively preempts whatever he was going to say. Tick quickly retreats.

In that letter was the poem you gave me as a little girl," I complete his unfinished thought.

"Yes," Grandpapa confirms. "She also included a final request, the only thing she ever asked of me, to look after her daughter and keep her safe."

All these wicked promises. "Why didn't she ask her husband to do that? Why did she have to involve you?" I vent my rage at this long-dead woman who stole my grandpapa's soul.

"Her husband was planning on defecting and taking their daughter with him. She knew his chances of success were nonexistent without the help I could provide as a KGB officer."

I glance at Grandpapa's hands which seem to be moving of their own accord to knot themselves together. "That's why the Adámas was never caught, why you didn't want anyone to mention his name and why you had some of the stolen Silver Bear diamonds," I pronounce, piecing it together.

"Yes, having some of the diamonds occasionally show up in misleading places allowed me to lead the investigation away from their escape route. You're extremely perceptive, Tatiana. I shouldn't be surprised, but I am." He eyes me with a keen respect, then reaches over to squeeze my hand. His grip is still firm, even though his skin is wrinkled and weathered with age spots. *The respect is going the opposite direction for me, Grandpapa. I'm losing all of it for you.* "So, the Adámas stole millions from the Mir mine, you purposely let him escape, planted some stolen diamonds with *me* and later bought the mine so you could cover your tracks, all the while putting your real family in danger for a woman who refused to marry you."

"Tatiana, it's not—"

"I'm not finished yet! Jumping ahead a few decades, both the Adámas and his daughter are dead, and I bizarrely stumble upon the one person in this world besides you who still knows about the Adámas and the stolen diamonds. *Voilà!* You magically appear here on campus and go straight to Joe in the hopes of keeping all this hidden and recovering the remainder of the diamonds. To ensure I won't find out about it, you order Joe not to tell me anything." *Gavno!* All these divisive secrets and promises, when will they end? Best to end them now. "I'm also perceptive enough to figure out that the only reason you took such huge risks with both your career and my life, in order to protect Syntyche's daughter, is because she's your daughter too. You don't have to deny that anymore," I flatly state.

He looks at me with confusion.

"Syntyche and the daughter you had together are both dead. No one ever found out about the Adámas and he's dead as well. Joe would never go against a promise he made, and he promised Kluge he would never speak of the diamonds. So, the only person's life you should still be meddling in is your other granddaughter's, Lina. If you agree to stay out of my life, I won't tell anyone about this whole mess."

"Joseph has been telling you far more than I thought he would, but he didn't tell you everything. Ellie, or Lina, as Joseph refers to her, is not my granddaughter." His steel-gray eyes are clear and sharp.

Is he outright lying to me? How can I tell?

"Joseph loves you with his whole heart and soul, Tatiana. Don't push him away."

"You have no idea what Joe feels toward me!" Joe doesn't even know what he feels toward me. A bottled-up rage detonates inside me. My grandpapa spends a few moments talking with Joe, and now he thinks he's an expert. Saying Joe loves me with his whole heart is a very idealized version of my "relationship" with him.

"I've been keeping tabs on Joseph for most of his life, Tatiana, so I know how he reacts. When he walked in here with me this morning, I was admittedly concerned the only reason he agreed to speak with me was because he hoped I could lead him to Lina, which I wasn't going to do. You need him so much more than she does. Using every ploy in the book, I tried to convince him that I would take care of her, that he no longer was bound to her by an oath of allegiance. After realizing I couldn't sway him, I threatened him that—"

"You blackmailed Joe!"

"No, Tatiana, it wasn't like that. I only pointed out that if he ever exposed my secrets to anyone, he'd never find Lina. Without any hesitation, he walked out on me, went straight to you and told all. He gave the ultimate sacrifice for you," my grandpapa forcefully insists.

"Do you any idea of what you've done to him? How could you have manipulated him so?"

"You're important to me, Tatiana, and I don't want to lose you the way I lost your mother. Joseph can help you in a way that no one else can. You need to see that," he pleads.

I feel like I hardly know who my grandpapa is anymore. He's morphed into a conniving, doting, old man, or perhaps I should say monster. Furious with him, I get up and march out the door of his life. Permanently.

Chapter 50

Self-pity is our worst enemy, and if we yield to it, we can never do anything good in the world ... Face your deficiencies and acknowledge them; but do not let them master you. Let them teach you patience, sweetness and insight.

Helen Keller

Joe
Monday, October 15

Anna disappears into the Skinner Memorial Chapel, and I stand by and wait. After her guard comes back out, I remain a little longer. How much will her grandfather actually tell her, and how will Anna react? Their chat could take quite a while, especially if he tells her everything he told me. Anna will have way more questions to ask him than I did. No use hanging around here.

A small part of me still wants to rush off and find Lina, but I've started to accept some of what Mr. Alliluyev was piling on me, predominantly the part about how Anna needs me in a special way, a way in which no one else can help her. So I head to my room to "rest up" from a concussion. Anna will know where to find me.

"Where have you been now, Joe, saving Anna and the world again?" Lance assails me the moment I enter. "Why do I even bother? Well,

I'm not going to anymore!"

He's even angrier than yesterday. "I had something I needed to do Lance," I feebly try to placate him.

"Like what, Joe? What's more important than recovering from a concussion? ... No answer to that? You've just blatantly proved your head injury is a sham. I've had it with all your stunts and I'm done covering for you. Asking me to give that fake doctor's note to the coaching staff was the last straw. Since you seem so hell-bent on getting kicked off the team and out of school, have at it. I'm out of here." He slams the door so hard it knocks his family picture off the wall, shattering the glass.

I drag the garbage can over and start picking up the larger pieces of broken glass. Chalk up one more notch for situations I've mishandled in my life. Hopefully, someday, I'll be able to explain all this to Lance. Falling on my bed, I attempt to rest. My endeavor is instantly stymied when Anna's supercharged emotions burst into my head a few minutes prior to her madly banging on the door.

Figuring out it's unlocked, she explodes into the room, spraying fragments of destructive energy in every direction. "You only came chasing after me because my grandpapa didn't tell you where Lina is," she viciously lashes out.

What the hell did he tell her? I sit up on the edge of the bed. "No, Anna," I frustratingly argue. "That's not true."

"If he had told you where Lina is, you'd be there right now," she pronounces, her face flushed with fury.

"700 Faith Way, Apartment 373, Minneapolis," I state.

She's momentarily stunned into silence, disoriented by the conflicting accounts. A lifetime devoid of emotions is hindering her comprehension of how feelings impact people's decision-making. Then her reset button activates and more accusations pour out of her mouth. "My grandpapa *never* would have told you where Lina is after you refused his demands. You just made that address up."

"He did tell me, Anna," I repeat in a slow, measured tone. "Your grandfather has different sorts of reasons for doing things. He wanted to force my hand, and in his own warped way, I think he was looking for forgiveness."

"Forgiveness for what? And if he did tell you, then why are you still here?" Her distrust of both him and me are wrapped around her questions.

"Forgiveness for not following through on his promises." That's my best supposition. "And I'm still here because your grandfather forced me to see that I can't love two people with all my heart and soul at the same time. By telling me where Lina is, he very effectively cornered me into making a choice." I walk over to within a hair's breadth of Anna. "You're the one I sought out, the one I love with all my being."

"But you still love Lina. You loved her first."

How can I explain this to her? Racking my brain, I hear Lina—Einstein's Lina. The vibrating energy resonating from her strings connects with me. *"From the past, will come the music, that will ring so true and free. In the madness of your torment, you will hear the melody. It will bridge the deepest chasms and will cross the wildest streams, it will reach into the cosmos and fulfill the long-lost dreams."* I sing my soul to Anna with these lyrics.

She's staring at me through a stream of tears that are tangled with pain. "You're singing that for Lina, Joe, not me."

"Maybe a little, but not the way you think. Lina helped me become the person I am now. We both strengthened each other enough to survive the tragedies in our lives, and because of that, I love her. But it's not at all the way I love you." My heart is racing with a desperate vulnerability. "You know me as no one else does and you understand me as no one else can." *Please, don't push me away, Anna.*

"I ... I love you, Joe," she whispers back, wholeheartedly breaking open our vulnerabilities. The vulnerability of feeling, the vulnerability of being misunderstood, the vulnerability of exposing our weaknesses, and most of all, the vulnerability of loving.

"Do you know what the name Syntyche means, Anna?" I ask. "It originated in the Greek language."

She smiles, because she understands me. "What does the name Syntyche mean, Joe?"

"It means joint destiny."

Chapter 51

Love is something eternal.
The aspect may change, but not the essence.
Vincent Van Gogh

Tatiana
Monday, October 15

The air around us is quivering with a strange energy. The power of it almost frightens me.

Bending his head to mine, Joe slides his heated lips along my skin, sweetly and reverently covering my trail of tears with his simmering kisses. Taking a step back, his face shines with such hope and joy that it almost starts me crying again. "There's no guilt about kissing you anymore, Anna, only pure ecstasy." He gives me an impish wink.

At long last, after constantly fighting this bond and denying our love, we embrace it in its entirety. Forgiveness floods through us for the present, the past and the future. As we revel in this delicious feeling of acceptance, an expression of baffled serenity crosses Joe's features, making his mouth crook up slightly into a lighthearted grin. Then he smiles at me, a full-fledged, gorgeous, delightful and dangerously sexy smile.

I reach out and run my fingers along his rough jawline. He hasn't shaved for a few days and the effect is distinctly appealing. Wrapping my arms around his waist, I unabashedly press my body to his, igniting a heated hunger that's hotter than sin. The alluring passion pulses wildly between us, and I kiss him deeply and thoroughly.

Responding to my flood of erotically charged emotions, Joe slowly glides his fingers around my neck to trace his fingertips over my lips. The streak of singed skin he leaves behind feels like it's on fire. Aching for more, I crush myself harder against him. Instantaneously, he reacts to my explicit movement by grinding his groin against mine.

All my senses become keenly awakened, resonating with a powerful intensity to create a sensual symphony of absolute pleasure: the firm pressure of Joe's mouth on mine, the slightly salty taste of his tongue, the smoldering warmth of his breath on my face, the titillatingly exquisite rumble of his deep, low moan, the wickedly exciting grasp of his hand on my butt and the blazing heat of his body pressed tightly to me. Each one of these sensations stokes the flame, spreading it with shocking speed to every nerve ending in my body. I've never experienced anything so incredible in all my life!

I want more, so much more. Physically trembling, I start fumbling to unbutton his shirt, causing a mischievously beguiling sparkle to dance in his Siberian blue eyes. Achieving my goal, I stroke his bare chest, gradually swirling my hands downward while watching the muscles around his stomach tense.

Joe gently untangles himself from me, locks the dorm room door and wedges a chair under the knob for added insurance. Letting out a sly chuckle, he returns and expectantly waits for me to resume where I left off. I carefully remove his shirt, making sure not to bump his injured hand, then I run my tongue up and down his upper body with a drunken giddiness.

Incapable of holding back any longer, Joe starts to pull off my shirt one-handedly. At his current reduced rate of speed, this is going to take forever, which at the moment is any amount of time more than five seconds. Unable to endure his sluggish progress, I finish the task myself and start on my bra.

"This I can do." Joe stops me, skillfully unhooks it, pushes the straps off my shoulders and sends it falling to the floor. His eyes rake over my bare breasts, watching their rise and fall quicken as my breathing accelerates. Soaking in the view, he enticingly traces circles around my nipples with his tongue, then skims his teeth over the tips before finally closing his mouth and suckling.

Ecstasy, pure ecstasy! For some magnificent moments, my breasts receive gloriously orchestrated minuets of his attention. When he pulls back, his gaze shifts to my pants, silently signaling. I hastily unfasten the clasp, slide them to the floor, then wiggle out of my underwear. His fingers seductively circle downward until meeting my clitoris where he works his magic, creating a crazed tension that builds and builds until a glorious explosion sends unadulterated rapture to every iota of my body. From the top of my head down to the tips of my toes, my world spinning with euphoria. I revel in the experience, my first-ever emotionally laced orgasm. It's out of this universe!

Now it's my turn to tune in and turn on. With an electrifying anticipation running through me, I divest Joe of the rest of his clothing and stand back to enjoy the sight of his gorgeously masculine body. Time to rock his world. I begin by brushing my lips along the base of his throat. His pulse is pounding. Next I provocatively nibble, lick and kiss my way lower, past his lean, muscled abdomen to teasingly halt just beneath his belly button.

Valiantly, Joe struggles to maintains control, wanting to allow me time to explore this wonderfully new, uncharted territory. I seize the opportunity, for twenty heartbeats or thereabouts, until his body becomes so taut and strained that I don't think he can withstand any more of this delicious torture. Showing mercy, I stroke his full erection. A visceral shiver jolts through him, encouraging me to squeeze, taste and suck with wild abandon … until he abruptly stops me.

Scrambling us over to his bed, Joe pins me in place with his strikingly blue eyes and his one good hand. “I love you, Anna,” he softly pronounces in a husky yet incredibly tender voice. Finally, I'm able to take the full plunge and totally give myself up to another. “I love you, Joe,” I answer back.

Primal need takes over and we boldly claim each other. Fire meets fire. Our inflamed kisses, combined with sizzling skin-to-skin contact, fuels the inferno. Joe lifts his hips to enter me. Wrapping my legs around his, I match his rhythmic strokes as he presses deeper inside of me. Bursts of bliss erupt, and I firmly cling to him while the impassioned mania of our flexing bodies races us toward the ultimate erotic peak, the climactic *la petite mort*.

Tangled together, we're propelled into oblivion, like the apocalyptic genesis of two galaxies colliding, and we meet eternity. There's no past or future, there's no beginning or end, there's only this unexplainable mysterious entanglement. And within this miracle, body to body, mind to mind and soul to soul, we find ourselves.

We are the ones we've been searching for.

Chapter 52

Somewhere in the vast expanses of the universe, two bodies orbit each other. Pulled together by an unseen force, they become locked in a steadfast grip and finally collide. Their powerful union produces ripples that extend across the vast expanse. These vibrations pass through the opus of space-time, squeezing and stretching everything in their path and opening a new window on existence. They are Einstein's Entangled Messengers.

Joe
Monday, October 15

As I collapse next to Anna, her body melts into mine and she lets out a heartfelt sigh. Curling farther into me, she brushes her lips against my face while softly murmuring, "I love you, Joe."

Side by side we lie, sheltered inside an almost-mystical, emotional healing bubble, a blissful retreat from our screwed-up lives. Amazed, I marvel at what she's done, given me the absolution to live my life.

Rap, rap, rap! "Piss off!" I yell.

Rap, rap, rap ... Rap, rap, rap! The urgent knocking continues. "I have to talk to you both." Anna's grandfather's gravelly voice booms through the locked door.

Reluctantly, Anna and I separate, but not before securely placing these special moments in amongst our joint memories, to cherish again and again. In a somewhat trance-like state, we retrieve our scattered clothes and get dressed. After helping me button my shirt and jeans, she plops on the edge of my bed and waits for the *grand* entrance.

The instant I unlock the door, her grandfather barges through, almost breaking apart the chair I'm in the process of removing. "I've been followed here," he solemnly informs us as he locks the door behind him. His two bodyguards remain posted in the hallway. "It's too dangerous for you to stay at Carleton, Tatiana, you have to relocate. The Russian regime will use you to get to me and silence my political protests."

"I'm not leaving Carleton or Joe," Anna calmly states, placing her hands in her lap like an elegant lady of the Victorian era, totally out of character for her. Where is the wild beast that arises anytime anyone dictates what she should do?

"Joe will go with you, but you both need to leave quickly. It's not safe here." Her grandfather's voice is shaky with fear.

"And fleeing to somewhere else will be safer? For how long? Hiding here got me abducted and nearly killed before the first trimester is even half finished. I'm not going to spend the rest of my life running away." Anna doesn't raise her voice at all, only intertwines her fingers and fiercely clenches them together.

Mr. Alliluyev slides my desk chair over and sits directly in front of her. "You deserve a chance at happiness, Tatiana, and you won't get it here."

"I already have it here." Anna looks over at me and smiles. Her face is practically glowing. She's the happiest I've ever seen her.

"But you won't be able to keep it here. I'll help you make arrangements to—"

"Let her make her own decisions," I angrily cut in, experiencing firsthand why Anna complains so vehemently about other people controlling her life.

Mr. Alliluyev rockets out of the chair, knocking it over in the process. That piece of furniture isn't going to survive being around him much longer. I hope he pays for it if he busts it, because my budget sure can't.

"You of all people, Joseph, should understand why I can't walk away and leave her here in danger. My actions, my politics, my money and my decisions have led to this." Returning his attention to Anna, he implores her from a different angle. "I lost my daughter, your mother, well before her time, and I can't go through something like that again. Children shouldn't perish before their parents."

His plea has a chilling effect on Anna, and I hastily step between them. "Off the coast of the Swedish island Öland in the year 1564, the blasting from constant cannonball shots was deafening. For two days, ferocious fighting and toxic smoke attached to terrifying screams fouled the air." Getting right in Mr. Alliluyev's face, I battle him head on to foul his air. "Hundreds of soldiers clambered on board the blood encrusted decks of *Mars the Magnificent*. Suddenly, a violent explosion rocked the vessel. The heat from the flames made the seas around *Mars* froth like a witch's cauldron. The greatest warship of its time was dying."

"What the hell are you saying, Joseph?" Mr. Alliluyev roars.

"Why did all those soldiers forfeit their lives boarding a ship that obviously had only minutes left afloat? Because they were so damn caught up in their own shortsighted objectives, they lost sight of the imminent danger right in front of them." I can't explicitly detail it out for him any more than that.

"I don't have time for your nonsense." Pushing around me to Anna's side, he kneels and places a hand on each of her shoulders. "I have the money and resources to create new identities, giving the two of you a chance for a life together. Don't be headstrong and insist on remaining at Carleton. In the end you will lose, and then everyone I love dearly will be gone. Please, let me die knowing you are protected from my mistakes."

I'll be damned if I'll let him manipulate her the way I've been manipulated. "Anna doesn't owe you a bloody thing. Get out," I yell, pointing to the door.

Seamlessly slipping into his KGB officer persona, Mr. Alliluyev rises to loom before me. "Let me make your current situation clear to you, Joseph. Markow's blood in the basement of the homeless shelter was recently discovered. The evidence will lead the authorities directly to you, and you'll end up in prison, unless I intervene," he cuttingly rebuffs me. "How will you be able to protect Tatiana from there?"

He's bluffing. After everything he's done to force me into her life, he's not going to rip me away from her. However, I have an ace up my sleeve too. "On February 28, 1926, Svetlana Stalina, the only daughter of the brutal Soviet tyrant Josef Stalin, was born," I coldly state. "When she was six years old, her mother shot herself through the heart, committing suicide to escape the fierce, overbearing control of Josef Stalin. Little Svetlana was never told the truth, and when she was a teenager, she read about her mother's suicide in an English journal. The shocking revelation destroyed a part of her. Hidden secrets have a way of causing devastating, unforeseen consequences, Mr. Alliluyev."

"Secrets also save lives, Joseph. They can save yours," he shrewdly duels with me.

"When Svetlana was seventeen, she fell in love," I continue. "But her father didn't approve and banished her lover to a labor camp in Siberia. In 1967, after three more romances and two failed marriages, she defected to the United States, where on November 22, 2011, she died alone in Richland County, Wisconsin. That's only a few hours from here." My words are filled with secondary implications and insinuations. I can tell Mr. Alliluyev is still keeping part of their family history hidden from Anna, and I'm fairly certain it has something to do with her mother's death.

"I don't need a history lesson from you," he angrily grinds out.

If Anna weren't sitting here, I doubt he would be this courteous. "I think you do, Mr. Alliluyev. To jar your conveniently spotty sharing of the circumstances surrounding Anna's mother's death, I'll recite more of Svetlana's story line for you to aid your memory. After Svetlana fled her Russian homeland, she was constantly moving. India, Switzerland, America, England, France, even back to Russia and Soviet Georgia for a bit. She was always searching for something she never found, peace of mind. Svetlana hated her past, stating her father left a shadow over her from which she could never emerge. 'Wherever I go, I will always be a political prisoner of my father's name,' You've put Anna in that same confining cage." I'm determined to force him to realize what his actions have already done to Anna and what they may do to her in the future.

"How dare you compare me to Stalin!"

Excellent, I've got him against the ropes, time for the finishing blow. "The parallels between Stalin's life and your life are strangely similar in many ways, and Anna puts a great deal of credence in coincidences. After Stalin died, Svetlana took her mother's last name, Alliluyeva, the feminine version of the surname Alliluyev."

Anna gasps in surprise and her grandfather's face flushes red. "That has absolutely no connection to me and zero bearing on Tatiana's life. You are trying to—"

"Stop!" Anna cries out, collapsing forward and grasping her head. In a heartbeat, we're both by her side, her grandfather patting her knee and me holding her hand.

Damn it. I did exactly what I accused Mr. Alliluyev of doing, getting so caught up in the battle that I lose track of what I'm fighting for. I'm no better than him. Contritely, I wrap my good arm protectively around Anna's waist, and she leans into me resting her head on my shoulder. I willingly share the burden of her pain.

After the headache eases some, she looks up at her grandfather. "I want to know everything about my mother. No more secrets. Is she related to Stalin? Why did she get these headaches? What was she so afraid would happen to me?"

Struggling to his feet, Mr. Alliluyev rights the knocked-over chair and drops down onto it. The chair creaks in complaint. I give it a lifespan of about thirty more minutes. "Katya was only twenty when the Chernobyl Nuclear Power Plant reactor exploded," he begins, his worries about Anna's headaches temporarily overriding his other concerns.

A shivers ripples through Anna, and I hold her tighter. "That photograph of Chernobyl, the one my mama earmarked in a book I found in her closet ... she was there, my mama was there when it exploded, wasn't she?"

Her grandfather recoils, still deeply affected by the disaster and its aftermath. "She was visiting her grandmother in the forests outside of Pripyat," he cautiously answers.

"Shit! That town is only six miles from the plant," I exclaim. "The radioactive fallout there was roughly four hundred times greater than the fallout from the Hiroshima bomb, and the Russian government didn't evacuate residents until a day and a half after the explosion." *Crap!* I shouldn't have said that.

Mr. Alliluyev scowls at me, warning me to keep my mouth shut. "Your mother started getting severe headaches and uncontrollable fits of coughing. Ultimately, she developed thyroid cancer."

"But she didn't still have cancer when I was alive, did she?" Anna questions.

Mr. Alliluyev collects his thoughts before responding. "She had surgery to remove her thyroid, followed by radioactive iodine treatment. It was successful, but her headaches never completely went away."

"So, the nuclear explosion was the cause of her headaches?" Anna presses him for a clearer answer, yearning to make sense of her own headaches.

"I don't know. The doctors all told me the same thing, she was suffering from migraines." Mr. Alliluyev recites this diagnosis with a sense of finality, yet his body language, along with the slight edge in his voice, betray him. There's more to it.

"Anna needs to hear the entire truth, Mr. Alliluyev."

He glowers his steel-gray eyes at me while attempting to determine what, if anything, I know, or if I'm bluffing.

"I don't want to die from these migraines, Grandpapa. Tell me!"

"Your mother ... she was always different, even as a young child," he hesitatingly admits. "She was aloof and unsociable, with broad mood swings that fluctuated at the drop of a hat. One moment she'd be happy, the next miserable, and you never knew what would set her off. A baby crying, an area with large crowds of people or a picture in a book."

Whoa! Anna seems to have inherited many of the traits Mr. Alliluyev just listed.

"Kee-eee, Kee-eeeeeee!" Two plaintive, high-pitched bursts of whistling pierces our ears. The first is a short two-second occurrence, followed quickly by a long four-second one. We all instantly swivel around. A small hawk is sticking out of the opening of a small house and sickly flapping its remaining wing. The front side of the contraption is painted a puke-lime green and the pendulum hanging from it is shaped like a neon-orange flower blossom. It's hideously obnoxious.

How is that thing still working? I thought for sure the boot I hurled at it earlier would have done the trick. This miserable contraption is one of the ways Lance is retaliating for my recent behavior toward him. Shortly after it went off the first time, he texted me that the noise was sure to help with my concussion. When I texted him my disagreement, he messaged back that my usage of the cell phone would help as well. If he doesn't come round soon, it's going to be a really long sophomore year.

"Kee-eeeeeee!" I throw the nearest book at the damn thing, smashing the bird off its perch, but the clock spurts out a few more raucous squawks. Anna crushes her hands over her ears, and Mr. Alliluyev goes over to remove the batteries. Sitting back down, nonplussed by the racket, he resumes his narration. "In addition to loud noises, your mother hated rules and would constantly complain that she was never allowed to *feel* free. The only things she found comfort in were her music tapes and our cats."

That description perfectly jibes with my limited knowledge of Anna.

"Did she ever tell you she could feel other people's emotions, Grandpapa?"

Mr. Alliluyev cringes and rapidly looks away. "She was always searching for explanations. How does the brain process emotions? Why do some people feel things so intensely? Why is it so difficult to understand the cause of feelings?" He chooses his next words very judiciously. "She claimed she was an empath."

"You didn't believe her." Anna looks like she was punched in the gut.

"No, yes, I don't know." Mr. Alliluyev's extremely uncomfortable. He's used to being in charge of situations, and this one is going in a direction he doesn't want it to. "She attempted to describe it to me when she was a teenager, claiming she not only felt the pain of those around her, she *lived* it."

"The pain of hundreds, yet the pain of one. I feel it sometimes too," Anna softly laments, commiserating with her mother as if she were sitting here next to her.

Now it's Mr. Alliluyev's turn to look like he was punched in the gut. I bet his daughter used that exact same phrase to describe her pain. "Do you have any water, Joseph?" He's becoming deathly pale.

I grab one of Lance's chilled bottles out of the mini-fridge and hand it to him. After a few small sips, he resumes talking. "The older she got, the worse the headaches got, until finally they dominated her life. She coped by isolating herself. No more watching television, listening to the radio, reading newspapers, going into town, and eventually, no more attending school. Katya incessantly argued that she had to get away from the city, that she needed even more isolation. After her eighteenth birthday, we gave in and let her go stay with her grandmother in the forests outside of Pripyat ... I'll never forgive myself for that decision."

"Was she able to control her feelings there?" Anna anxiously asks, hanging on his every word.

"I think she may have learned to exist with them at times," he gruffly replies.

"Under what circumstances and for how long?" Anna's determined to pry every gory detail out of him and is getting frustrated with his vague answers. "What methods did she use and where did she learn them?"

"It's best left in the past, Tatiana. I don't want you to suffer the way she did." Mr. Alliluyev is visibly aging as he relives this for her.

"It's too late for that, Grandpapa. I started getting my mama's deadly migraines in September. The pain of hundreds, yet the pain of one is inside of me too, and I can't control it."

Mr. Alliluyev's face turns to ashen gray, rattled to the core by Anna's revelation. Even though this blatant fact has been staring him in the face this entire conversation, he's refused to acknowledge it, because then all doubt would be gone and all hope that she had avoided her mother's fate would vanish. Schrödinger's cat can be a nasty, feral beast at times.

"She would attempt to calm her mind," he numbly mumbles. "She read somewhere that the energy of the emotions around us must be allowed to flow freely, otherwise they'd upset the natural balance."

"Wu wei," Anna incredulously exclaims.

"If that's what you want to call it. I say it's pure nonsense, balancing our lives by doing nothing. It's preposterous, harmful psychobabble used to hoodwink vulnerable, confused and susceptible young girls," her grandfather angrily snaps.

"You just said it sometimes helped my mama exist with her intense feelings," Anna points out the contradiction.

"I said it was one of the things she tried, not that it helped. Sheer power of will was how she dealt with the migraines, and that was what allowed her to function in this world again," he forcefully asserts, frustration mapped over his features.

I believe in one thing only, the power of human will. Mr. Alliluyev is touting the same tenet as Stalin. If Anna weren't in such a vulnerable state right now, I'd shove that "coincidence" down his throat.

"Why were my mama's migraines so bad again later in her life, when I was alive?" Anna's tensed up, fearing yet needing to hear the reason.

"Your mother was headstrong and stubborn, and so is your father." Mr. Alliluyev's getting stretched to the breaking point.

"That's not an answer," Anna shouts, then balls up in pain. Her head is pounding viciously, therefore, so is mine. It's getting extremely hard to think straight. *I'm suffering the effects of a concussion, Lance.*

Payback is a bitch. I lower my head and try to ease some of the agony.

"I'm sorry I said that, Tatiana. Your mother loved you very much and since the day you were born she did everything in her power to protect you … because it was after your birth that her severe migraines returned."

"Protect me from what?" Anna pleads.

"Her huge mood swings, her depression, her pain. You reacted very dramatically to her; if she was sad, you were devastated, if she was afraid, you were petrified, and if she cried, you wailed hysterically and became inconsolable." His voice drops so low we can barely make out his next words. "Katya was terrified you'd get the migraines and be forced into the tormented existence she had lived in for so long."

"Then she *did* figure out how to block emotions!" Anna's body goes rigid. "And she locked them inside of herself, even though she needed to let them flow free to survive. They grew stronger and more powerful and … she died protecting me. She died because of me!"

"No!" Mr. Alliluyev frantically clutches her. "Your mother was not an empath, Tatiana, there's no such thing."

"No one believed her," Anna soulfully grieves.

"I believe," I earnestly profess. In spite of the massive volumes of contradictory, factual evidence my rational mind is throwing at me, I believe.

Chapter 53

"What is music at the end?
It is a movement.
Who is moving? Is it the sound?
No, it is the consciousness of who is listening to
and of who is moving the sound."
Sergiu Celibidache

Tatiana
Monday, October 15

A deafening silence crowds into the room as I sit there immobilized with guilt. "She died because of me," I wretchedly repeat.

"No, Tatiana. She died for you, not because of you. If there is blame, it lies with your father. He abandoned her when she needed him most, refusing to see what was happening. I pleaded with him to help her, but he ignored me." Grandpapa's bitterness sours the air, and the accompanying acerbic taste in my mouth makes me feel nauseated.

"But Father said he took Mama to all the best doctors, and they all told him the same thing, that she had migraines and needed rest." I find myself in the strange position of defending my father.

"He talked to you about this?" Grandpapa's shocked. Until very recently, my father refused to speak about my mama, especially with me. So through the years, I've continually badgered Grandpapa to tell me about her. Even after he fled to England and we communicated mainly through written messages that his associates slipped to me before the start of my combat Sambo lessons—which my father insisted on—I always had questions about her at the ready. Most of the time, he never gave me any answers about my mama's headaches, or at least not with a reply that referred to what I really wanted to know. What caused them? It seems on that one front, hiding this information from me, Grandpapa and Father were co-conspirators.

"Father talked with me about the headaches just after the kidnapping," I absentmindedly murmur to him, letting my mind wander within memories of my mama. I'm four years old, sitting at the kitchen table. Mama's face is its usual pale color, and her hair is pulled back in a braid. She's in the middle of fixing eggs for breakfast when suddenly she drops the spatula, turns so I can't see her face anymore and grabs her head. After a short moment, she glances at me with a small smile, and the pain inside my head goes away. I didn't know back then it was the other way around; I was experiencing her pain, not her experiencing mine. The suffering of hundreds, yet the suffering of one.

Joe grips my hand and squeezes, returning me to the here and now, to the perspective of a twenty-one-year-old, With the additional knowledge about my mama, I reflect on what happened to her. Yet no matter how many ways I view it, I can't see that it was my father's fault. She died because of me.

"Ma, which stands for mega-annum—one million years—is a celestial mechanics term signifying very long periods of time into the past or future." Joe forces me to look into his eyes. "We are influenced by where we came from and where we will go to. Therefore, it could be said the human soul passes through a Ma of our ma's suffering. The pain travels with us."

Joe's words comfort me a little. So, I'm experiencing *my ma's* Ma of pain? I chuckle. Only he could come up with that obscure coincidence of connection. But there's no way I can deal with a million years of suffering! I have to focus on my more immediate issues. "Did she ever go back to Pripyat after the Chernobyl disaster?" I ask, needing a deeper understanding of this to have any chance at controlling the migraines.

"An exclusion zone was set up and public access was denied," Grandpapa states, not actually answering my question. It's a skill both he and Joe are very adept at.

"The exclusion zone covers over one thousand miles, Mr. Alliluyev. The area directly around Pripyat, which is strictly off-limits, is called the Zone of Alienation," Joe corrects him. "However, that's not what Anna asked. How often did your daughter return there?"

Grandpapa remains silent, his lips stretched thin and tightly pressed together.

"Scientists estimate that area won't be safe for human life for another twenty thousand years," Joe keeps on. "Pripyat is an abandoned ghost town where almost everything was left behind. Personal belongings are strewn in the streets and pictures still hang on the walls. It's a place frozen in time, just like Pyramiden."

Pyramiden! The deserted schoolyard forms in my mind, rusted slides and idle swings dangling eerily still, emanating desolation and abandonment. The similarities between Pyramiden and Pripyat, the longings for a life left behind, are unsettling and agitating. Distantly. I hear Grandpapa yelling at Joe, but I'm too deep in the rabbit hole to discern what he's saying. The anguish within me is reaching a dangerous level. I'm doing my best to let it flow through, yet it's not enough.

"Anna!" Joe joins me wholly in the misery and helps me find my way back. Cracking open my eyes, I notice I'm looking up at the ceiling.

Grandpapa's panicked face floats over me as he props a pillow under my head. Simultaneously, he's hollering at Joe. "Why did you upset her like that?"

"Because keeping secrets from Anna isn't protecting her, it's hurting her. She needs to know everything about what happened to her mother, or she'll never be able to manage the grief. Now tell us why your daughter keep going back to Pripyat, Mr. Alliluyev."

Grandpapa fixes his gaze on me. "Are you sure you're ready to hear this, Tatiana?"

"Yes," I answer without any hesitation. I've been waiting my whole life to hear this.

"She was searching for something she desperately wanted to find, release from the torment that kept breaking her down and pulling her back." A sob catches in his throat.

The horrendous heartache, the unbearable loss and loneliness that plagued my mother is physically palpable to me. These emotions, which were formed before I was born, created in a place I've never been to and linked to a soul-wrenching trauma I know nothing of, are inside of me, an integrated part of me. Slowly, I wade into them with a wavering resolve, preparing myself for whatever my grandfather is about to disclose.

Next to me, Joe is clenching his teeth and sucking in short spurts of breath in order to cope with the onslaught I'm sending his way. "It was recently discovered ... that the psychological effects of great trauma ... are genetically passed on to future generations," he rasps. "Anna *needs* to know ... where did your daughter go ... when she returned to Pripyat?"

"The cemetery," Grandpapa chokes out, his voice sunken and hollow. "No matter what I said or did, I couldn't deter her, so I gave her protective gear to shield her from the radiation."

"Who's grave did she go to see?" I ask. "Who did she risk her life for?"

"Don't ask any more questions, Tatiana. It will consume you the way it consumed your mother."

"It's already consuming me!" I bolt upright on the bed and glare at him.

"Shall I start the list of tombstones for you, Mr. Alliluyev?" Joe breaks in. "Victims of the massacre enacted after the October Revolution, casualties of the great fire, innocent people executed on that very spot during World War II, and more recently, fatalities of the Chernobyl disaster. It's likely many of the deceased buried there are ancestors of Anna's, dating back hundreds of years. The grief of hundreds." Joe pauses, straining against the horrific pain radiating from me. "There's also someone very special to Anna's mother there, someone she mourned her entire life. Anna and I can both feel it, the grief of one."

Torrents of tears are ravaging my grandpapa. His heart has been viscerally slashed. "She went to a makeshift tombstone ... with a lone rose engraved on it. Her name was Rose ... she was your sister, Tatiana."

Sister! Shock, anguish, grief. The grief of one. *I had a sister!* Surprise, curiosity, connection. A spark lights in my heart. *I have a sister!* I want to know everything about her. "What color hair did Rose have? What color eyes? What was she like?" I excitedly pump Grandpapa for information.

"Her hair was jet-black, her eyes were deep brown and she was just like you ... for the eighteen, joyously bittersweet months of her short life." Grandpapa's eyes are bottomless wells of sorrow, rippling with regrets.

"Rose," I reverently test the sound of her name. "Rose," I speak louder, enjoying how it makes me feel. "Who was Rose's father, is he still alive and why did no one ever tell me about her?"

"Fyodor Igorevich Alexeev, a local boy who worked at the nuclear plant. He died soon after the explosion from radiation poisoning." Taking out his handkerchief, Grandpapa noisily blows his nose, then shoves the hankie back in his pocket. "Rose was the loop to your mother's soul, Tatiana, and after she died, your mother became a lifeless shell, consumed by grief. When you were born…" Grandpapa nervously rubs his hands together.

When I was born, what? What happened then? Should I ask him right away or give him a moment?

"In mathematics," Joe cuts in, "a rose is a plane curve consisting of three or more loops meeting at the origin and shaped like a petaled flower. It is the only wave form that retains its shape when added to another wave of the same frequency."

"Rose, Rose, Rose," I lilt her name in different pitches, feeling out the different frequencies, then sit quietly listening to them in my head. I can hear Rose's cries because they're just like the sound of my cries, and I can hear her laughing because it's just like my laugh. Is she "the one"? Is Rose the person the old woman in my dream about Mars told me I would make all the difference for? *Get a grip Tatiana, Rose is dead and you can't make a difference for her now.*

"In Greek mythology, roses signify immortal love that withstands time and death," Joe recites in an odd voice. His current manner of dealing with my wild shifts of emotions, following the word "Rose" wherever his mind leads him, is soothing. "In tarot, a rose is considered a symbol of balance that expresses hope and new beginnings."

New beginnings! The tangible world fades away, and Joe and I are transported by our special, entangled string of existence. Amongst the wondrous beauty of twinkling energy, I hear it clearly, the music of the stars, the rich melody of timeless cosmic song which lifts our consciousness. Celebrating this astonishing bond, which I once abhorred but now cherish, I feel the connections down to the origins of my being.

Rap, rap, rap! "Mr. Alliluyev, the pilot is waiting for your answer."

"I gave strict instructions not to interrupt me," Grandpapa booms at the locked door, plummeting Joe and me back into this tiny dorm room.

Misunderstanding my strange reactions to his revelations, Grandpapa tries to calm me. "Roses also symbolize sacrifice, Tatiana, and that's exactly what your mother did for you. The reason you never learned about Rose is your father forbade it, demanding she be left in the past. After your parents were married, Katya was never allowed to mention her name. The immense sadness and heartache she locked in her heart slowly destroyed her. It was your father's rigid, stubborn insistence that killed your mother, not you."

Why would he do that? How could Father demand that of my mama?

"I can't bear to lose you, Tatiana. Let me move you somewhere safe," Grandpapa pleads.

"Is relocating Anna what's best for her or for you, Mr. Alliluyev?" Joe angrily confronts him.

Rap, rap, rap! "The pilot needs an answer now if you want to take off on time. What should he put on the flight plan?" a bodyguard courageously shouts through the door.

Grandpapa glances back and forth several times between Joe and me. The fight is gradually slipping out of him, and in its place, a defensive cloak of fatalism is covering his hunched form. "Tell him I'll be flying back to England. Alone." With a herculean effort, he slowly hauls himself out of the chair, and I rise to meet him. Trudging forward a step, his foot dragging across the floor, he passionately enwraps me in a big bear hug.

Bear hug ... Water bears ... Tatty. All things that will survive the end of life as we currently know it.

"Water bears can withstand extreme conditions," Joe pensively voices, causing Grandpapa's brows to furrow. Releasing me, he twists his head toward Joe.

"Water bear, Grandpapa. Remember, *Tatty,* my pet bear." I emphasize her name to jog his memory. "In his unique way, Joe is telling you that he understands, and he'll always be here for me, holding me strong."

A conciliatory look passes between them, then Grandpapa gives a slight bow of his head in acknowledgment. "Take care of her, Joseph." With heavy footfalls, he plods to the door and disappears from sight. I wish I had known it would be the last time I'd ever see him.

"What happens now?" Joe questions.

"I don't know." There's so much more for me to discover, about Rose, about my mama and about myself.

"You're not going to Pripyat, are you?" he anxiously asks.

"No." There's a deep vibration thrumming in the background beckoning me to a different place where there are songs without words yet the music's still heard. It's been singing to me all my life, but it's not summoning me to Pripyat or back to Russia or London, or anywhere else on this planet. It's calling me to venture to the stars, and I'm going to follow it all the way to Mars.

"This is my journey," I whisper. "Out where the stars live, where my spirit belongs, free to sing all its songs. This is my journey, and one way or another, Joe's coming with me!"

The end of Book 1 in the Entanglement Series.
Or is it the beginning?

Interesting snippets

Mr. Einstein rarely traveled anywhere without his violin. While he owned several over his lifetime, all were given the same nickname "Lina," which is short for violin.

Ancient humans believed the celestial spheres orbiting Earth produced divine music by their movement. The Greek philosopher, Pythagoras, appears to be the first to quantitatively explain the nature of celestial bodies using musical tones. This mathematical philosophy is referred to as Musica Universalis or the music of the spheres.

Modern humans use the Kepler space telescope to 'listen' to the sounds of the stars. The science of astroseismology is done by sensing the resonances inside a star and measuring the micro-changes in their intensity.

Alexander Litvinenko, a former KGB spy, accused President Vladimir Putin of orchestrating his killing in a statement read after his death. Here is an excerpt from that statement.

But as I lie here I can distinctly hear the beating of wings of the angel of death. I may be able to give him the slip but I have to say my legs do not run as fast as I would like.

I think, therefore, that this may be the time to say one or two things to the person responsible for my present condition.

You may succeed in silencing me but that silence comes at a price. You have shown yourself to be as barbaric and ruthless as your most hostile critics have claimed.

You have shown yourself to have no respect for life, liberty or any civilized value. You have shown yourself to be unworthy of your office, to be unworthy of the trust of civilized men and women.

You may succeed in silencing one man but the howl of protest from around the world will reverberate, Mr. Putin, in your ears for the rest of your life.

May God forgive you for what you have done, not only to me but to beloved Russia and its people.

https://edition.cnn.com/2006/WORLD/europe/11/24/uk.spy.statement/

Did this novel connect with you? I'd like to hear your thoughts and opinions to help enhance my other books in this series. Please take a few minutes to review it on Amazon at https://www.amazon.com/review/create-review/?ie=UTF8&channel=glance-detail&asin=B0BRLVSHMD

About the Author

Suzanne Jantscher is a former Naval Intelligence Officer and a retired air traffic controller with a private pilot license, scuba diving certification and a never-ending sense of adventure. Her joy of backpacking has led her from the Superior Trail in her home state of Minnesota to the Heaphy Track in New Zealand, from the Inca Trail in Peru to the summit of Mount Snowdon in Wales. When not writing or hiking, she spends time gardening, penning poems, playing board games and volunteering. She has three adult children and resides with her husband, an amateur astronomer, her dog, Penny, and a lifetime collection of books.

You can connect with me on:

https://www.facebook.com/profile.php?id=100088920124829

Made in the USA
Monee, IL
27 February 2023